THEORY OF
CONTINUOUS
GROUPS

Mathematicians of Our Time

Gian-Carlo Rota, Editor

CHARLES LOEWNER

Theory of Continuous Groups

Notes by
Harley Flanders and Murray H. Protter

THE MIT PRESS

Cambridge, Massachusetts, and London, England

CONTENTS

PREFACE

Charles Loewner, Professor of Mathematics at Stanford University from 1950 until his death January 8, 1968, was a Visiting Professor at the University of California at Berkeley on five separate occasions. During his 1955 visit to Berkeley he gave a course on continuous groups, and his lectures were reproduced in the form of duplicated notes. Loewner planned to write a detailed book on continuous groups based on these lecture notes, but the project was still in the formative stage at the time of his death. Since the notes themselves have been out of print for several years and since the opportunity of having a book by Loewner on the subject has passed, we decided to revise and correct the original lecture notes and make them available in permanent form. We are grateful to the M.I.T. Press for its cooperation in this project.

Loewner became interested in continuous groups when he studied the three-volume work on transformation groups by Sophus Lie. Well known for his mild manner, Loewner was, nevertheless, rather caustic in describing the shortcomings of Lie's presentation. Because Lie's books contained numerous illustrative examples, Loewner managed to reconstruct a coherent development of the subject by synthesizing the examples, many of which appeared only as footnotes. Over the years Loewner retained a strong interest in continuous groups, particularly with respect to possible applications in geometry and analysis. The illustrative examples in these notes, mostly geometric in character, reflect the unique way in which Loewner viewed each of the topics he treated.

HARLEY FLANDERS
MURRAY H. PROTTER

July 1970

THEORY OF
CONTINUOUS
GROUPS

LECTURE I

1. Transformation Groups

Let $S = \{p, p_1, p_2, \cdots\}$, $T = \{q, q_1, q_2, \cdots\}$ be arbitrary point sets and α a mapping on S into T. We write $q = \alpha p$ when q is the image of p under α. If each point of T is the image of some point in S, we say the mapping is **onto.** If $U = \{r, r_1, r_2, \cdots\}$ is a third set and β is a mapping from T to U, then we form the **composition** $\beta\alpha$, a mapping of S into U, defined by $(\beta\alpha)p = \beta(\alpha p)$.

We consider in particular mappings of a set S into itself. In this case the composite of two mappings is always defined. If α, β, γ are three mappings of S into S, then we have the **associative law**

$$(\alpha\beta)\gamma = \alpha(\beta\gamma).$$

There is a unique mapping ϵ, called the **identity,** of S onto S, such that for each mapping α

$$\alpha\epsilon = \epsilon\alpha = \alpha.$$

It is defined by $\epsilon p = p$ for each p in S. If α is a one-one mapping of S onto S, there is a unique mapping α^{-1}, the **inverse** of α, such that

$$\alpha\alpha^{-1} = \alpha^{-1}\alpha = \epsilon.$$

If $\alpha p = q$, then $\alpha^{-1}q = p$.

We now restrict consideration to one-one mappings of S onto S. The totality of such mappings forms a class $\mathfrak{G}$, which is closed under the operations of composition and formation of the inverse. It is clear that $\mathfrak{G}$ is a **group.**

The simplest example of such a group occurs when S consists of a finite number of points. In this case the group of mappings is the set of all permutations of the points of S. Each permutation is a **transformation** of the elements of **S** onto **S**. Thus the collection of permutations forms a **transformation group.** The set of all even permutations of **S** is itself closed under the operations of composition and formation of the inverse, and thus is a subgroup of the group of all permutations. The study of such permutations originated in the theory of algebraic equations and led to the development of Galois theory.

Groups of mappings have also been studied in geometry and have led to the theory of **transformation groups.** We give several examples from geometry:

1. **S** is the straight line. The transformations are the translations $x' = x + a$, where a is real. If $x' = x + a$ and $x'' = x' + b$ are two translations, the rule for composition is $x'' = x + (a + b)$, another translation. The inverse of $x' = x + a$ is $x = x' + (-a)$.

2. **S** is the straight line. The transformations are the mappings $x' = ax + b$, $a \neq 0$. It is clear that these mappings form a group; those for which $a > 0$ form a subgroup.

3. **S** is the projective line. The transformations are the mappings

$$x' = \frac{ax + b}{cx + d}, \qquad \text{where} \quad \Delta = \begin{vmatrix} a & b \\ c & d \end{vmatrix} \neq 0.$$

In this case it is necessary to adjoin the point at infinity to **S** in order to preserve the one-one character of the mappings. The collection of mappings for which $\Delta > 0$ is the subgroup of **proper** projective transformations.

4. **S** is the plane. The transformations are the translations

$$\begin{cases} x' = x + a \\ y' = y + b. \end{cases}$$

5. **S** is the plane. The transformations are the **proper motions**

$$\begin{cases} x' = x \cos t - y \sin t + a \\ y' = x \sin t + y \cos t + b. \end{cases}$$

The **improper motions** are given by

$$\begin{cases} x' = x \cos t - y \sin t + a \\ y' = -x \sin t - y \cos t + b. \end{cases}$$

The totality of the proper and improper motions forms a group.

6. **S** is n-dimensional space. The transformations are

(*) $$x_i' = \sum_{k=1}^{n} a_{ik} x_k, \qquad i = 1, 2, \cdots, n, \quad \det |a_{ik}| \neq 0.$$

The non-homogeneous transformations, which also form a group, are

$$x_i' = \sum_{k=1}^{n} a_{ik}x_k + a_i, \qquad i = 1, 2, \cdots, n, \quad \det |a_{ik}| \neq 0.$$

Those transformations (*) which have the property

$$\sum_{i=1}^{n} (x_i)^2 = \sum_{i=1}^{n} (x_i')^2$$

form a subgroup, the **orthogonal group**. If A denotes the matrix $\|a_{ik}\|$, A' the transpose of A, and I the identity matrix, then the orthogonal group is characterized by the condition

$$AA' = I.$$

7. **S** is the complex z-plane. The mappings are the **proper Möbius transformations**

$$z' = \frac{az + b}{cz + d}, \qquad \begin{vmatrix} a & b \\ c & d \end{vmatrix} \neq 0, \quad a, b, c, d \text{ complex numbers.}$$

These transformations map circles (including straight lines) into circles. Here the point at infinity also must be included in **S** in order to preserve the one-one character of the transformations.

The **improper Möbius transformations** are the mappings

$$z' = \frac{a\bar{z} + b}{c\bar{z} + d}, \qquad \begin{vmatrix} a & b \\ c & d \end{vmatrix} \neq 0.$$

The totality of proper and improper Möbius transformations constitute the **inversion group**.

Let $\mathfrak{G} = \{\alpha, \beta, \gamma, \cdots\}$ be an abstract group with the operation of composition. The elements of $\mathfrak{G}$ satisfy

    a. Associative law: $(\alpha\beta)\gamma = \alpha(\beta\gamma)$ for all α, β, γ in $\mathfrak{G}$.
    b. Existence of a unique unit ϵ such that $\alpha = \epsilon\alpha = \alpha\epsilon$ for all α in $\mathfrak{G}$.
    c. Existence of a unique inverse α^{-1} for each α in $\mathfrak{G}$ such that $\alpha\alpha^{-1} = \alpha^{-1}\alpha = \epsilon$.

Two groups $\mathfrak{G}$ and $\mathfrak{G}' = \{\alpha', \beta', \gamma', \cdots\}$ are **isomorphic** if there is a one-one correspondence between the elements of $\mathfrak{G}$ and $\mathfrak{G}'$ so that if $\alpha \longleftrightarrow \alpha'$, $\beta \longleftrightarrow \beta'$ then $\alpha\beta \longleftrightarrow \alpha'\beta'$.

THEOREM 1.1 (A. Cayley). *Let $\mathfrak{G}$ be an abstract group. Then $\mathfrak{G}$ is isomorphic to a transformation group.*

Proof. Let α be a fixed element of $\mathfrak{G}$. Then for each ξ in $\mathfrak{G}$ we define a mapping of $\mathfrak{G}$ onto $\mathfrak{G}$:

$$\alpha\xi = \eta.$$

This mapping $\xi \longrightarrow \eta$ we denote by τ_α. The set $\{\tau_\alpha\}$ forms a transformation group, and the correspondence $\alpha \longrightarrow \tau_\alpha$ is an isomorphism. First, the relations $\alpha\xi = \eta$, $\beta\eta = \zeta$ imply that $\beta(\alpha\xi) = (\beta\alpha)\xi$, hence $\tau_\beta\tau_\alpha = \tau_{\beta\alpha}$. Next for $\alpha \neq \beta$ we see that $\alpha\epsilon \neq \beta\epsilon$, hence the transformations τ_α and τ_β must be different. The relation $\alpha\xi = \xi$ holds for all ξ only if $\alpha = \epsilon$. Thus the correspondence $\alpha \longleftrightarrow \tau_\alpha$ is one-one.

Let **S** be an arbitrary set and $\mathfrak{G}$ a transformation group of **S** onto **S**. A subset of **S** is called a **figure.** It is possible to construct new transformation groups from $\mathfrak{G}$, the new transformations acting on figures instead of points of **S**. For example, let a figure of **S** consist of a pair (p_1, p_2) of points of **S**. Let $\alpha p_1 = q_1$ and $\alpha p_2 = q_2$. These relations define the mapping $(p_1, p_2) \longrightarrow (q_1, q_2)$. If $\mathbf{S}_1 = \mathbf{S} \times \mathbf{S}$, then $\alpha(p_1, p_2) = (q_1, q_2)$ defines a group of transformations of $\mathbf{S}_1$ onto $\mathbf{S}_1$. Other possibilities for generating new groups occur if we take pairs (p_1, p_2) with the additional assumption that $p_1 \neq p_2$; also we may identify the pair (p_1, p_2) with (p_2, p_1). Other figures consisting of $3, 4, \cdots$ points, ordered, distinct, etc., yield further transformation groups.

Let **S** be an arbitrary set and $\{\mathbf{L}\}$ a decomposition of **S** into disjoint subsets. In such a case we say that **S** is partitioned into **layers.** Suppose $\mathfrak{G}$ is a transformation group of **S** onto **S** with the property that for any layer **L** of $\{\mathbf{L}\}$ and any α in $\mathfrak{G}$, the map of **L** under α is itself a layer. That is, $\alpha\mathbf{L} = \mathbf{L}'$. The layers themselves form a new space, and the group $\mathfrak{G}$ generates a new group acting on the space of layers; these layers may be considered as figures of **S**.

As an example let **S** be the plane and $\mathfrak{G}$ the group of translations

$$\begin{cases} x' = x + a \\ y' = y + b. \end{cases}$$

Divide **S** into the layers $\{\mathbf{L}\}$ consisting of the family of all lines parallel to a given line. It is clear that under the group of translations the layers map into layers, and that the induced transformations of lines into parallel lines form a group. We note that the original space **S** is two-dimensional, while the layer space $\{\mathbf{L}\}$ is one-dimensional. In general, such a decomposition is called a **covariant decomposition.** If the group $\mathfrak{G}$ is the set of all motions in the plane (Example 5, above), then this decomposition $\{\mathbf{L}\}$ does not generate a group, as layers do not transform into layers.

Let **S** and $\mathfrak{G}$ be as described. We call two points p and q **equivalent,**

$$p \sim q,$$

if there is a transformation α in $\mathfrak{G}$ such that $\alpha p = q$. This relation is clearly reflexive, symmetric, and transitive. Hence we obtain a decomposition of **S** into disjoint subsets, or layers. By this process the set **S** is decomposed into what are termed **invariant layers** with respect to $\mathfrak{G}$, since each layer is transformed into itself under the transformations of $\mathfrak{G}$.

The fundamental problem of invariant theory is the determination of the invariant layers of a space **S**. A general technique for finding these layers is the determination of functions defined on **S** which are constant on each invariant layer. That is, we seek one or more functions $F(p)$ such that $F(p) = F(q)$ if $p \sim q$. A **complete system of invariants** is a set of functions $F_1, F_2, \cdots, F_n$ such that $F_i(p) = F_i(q)$ for $i = 1, 2, \cdots, n$ *if and only if* $p \sim q$. If we have such a complete system of invariants, then the relations

$$F_1 = C_1, \quad F_2 = C_2, \cdots, \quad F_n = C_n,$$

where the C_i are constants, define an invariant layer.

As an example, we consider $\mathfrak{G}$, the group of rotations about the origin in three-dimensional space ($=$ **S**). An invariant layer consists of the surface of a sphere with center at the origin. A complete system of invariants consists of the single function r, the distance of a point from the origin.

A group $\mathfrak{G}$ of transformations of **S** onto **S** is **transitive**, if any two points of **S** can be transformed into each other by an element of $\mathfrak{G}$. That is, there is only one invariant layer. All other groups are **intransitive.** A group $\mathfrak{G}$ may be transitive, but some of the groups derived from $\mathfrak{G}$ (i.e., by considering figures) may be intransitive. As an example, the group $\mathfrak{G}$ of all motions in the plane is a transitive group. The group derived from $\mathfrak{G}$ by taking as figures pairs of points is intransitive. For a pair (p_1, p_2) can be mapped into another pair (q_1, q_2) only if dist$(p_1, p_2) = $ dist(q_1, q_2). A complete system of invariants for this group is given by the distance function between two points, dist(p_1, p_2).

The projective group on the line (Example 3, above) is a transitive group. In fact, the groups derived by taking figures of two, three, distinct points respectively are also transitive, as it is known from projective geometry that any three distinct points may be mapped into any three distinct points by a projective transformation. On the other hand, the group derived by considering figures of four distinct points is intransitive. It is known that a set of four distinct points can be mapped onto another set of four points projectively if and only if the cross-ratios of the sets are the same. Hence, a complete system of invariants for this derived group is the cross-ratio of four points. A problem of invariant theory is the determination of complete systems of invariants, not only for a given group $\mathfrak{G}$ of transformations of **S**, but for the groups derived by the consideration of figures of **S**.

Let $\mathfrak{G}$ be an abstract group. According to Cayley's theorem, Theorem 1.1, we can form the transformation group $\{\tau_\alpha\}$ isomorphic to $\mathfrak{G}$. The transformations τ_α: $\alpha\xi = \eta$ we call **left translations.** We obtain another group by taking **right translations:** $\xi\alpha = \eta$. In this case we again obtain a group of transformations which is in one-one correspondence with $\mathfrak{G}$ and satisfies the condition $\tau_{\alpha\beta} = \tau_\beta\tau_\alpha$. The mapping $\alpha \longrightarrow \tau_\alpha$: $\eta = \xi\alpha$ is an **anti-isomorphism,** and the group thus generated is anti-isomorphic to $\mathfrak{G}$. The right translations yield an isomorphism by taking the correspondence $\alpha \longrightarrow \tau_\alpha$: $\eta = \xi\alpha^{-1}$. Then the relation $\tau_{\alpha\beta} = \tau_\alpha\tau_\beta$ follows. In this case the space **S** is the space of the group itself.

For the Cayley representation of an abstract group $\mathfrak{G}$, we seek the most general covariant decomposition into layers. Suppose we have a decomposition into layers. Let $\mathfrak{H}$ be the layer containing the unit ϵ.

THEOREM 1.2. $\mathfrak{H}$ *is a subgroup of* $\mathfrak{G}$.

Proof. Let α be an element of $\mathfrak{H}$. Since $\mathfrak{H}$ is a layer, the elements of $\alpha\mathfrak{H}$ form a layer. But ϵ is in $\mathfrak{H}$ by hypothesis, and $\alpha\epsilon = \alpha$ is in $\mathfrak{H}$. Hence the layer $\alpha\mathfrak{H}$ is $\mathfrak{H}$ itself. Therefore, if β is any other element in $\mathfrak{H}$, then $\alpha\beta$ is in $\mathfrak{H}$. Since $\alpha^{-1}\alpha = \epsilon$ the layer $\alpha^{-1}\mathfrak{H}$ is $\mathfrak{H}$ itself, and $\alpha^{-1} = \alpha^{-1}\epsilon$ is in $\mathfrak{H}$. Thus $\mathfrak{H}$ is a subgroup.

Let $\mathfrak{H}$ be an arbitrary subgroup of a given abstract group $\mathfrak{G}$. Then $\mathfrak{H}$ generates a definite covariant decomposition of $\mathfrak{G}$ into layers. These are the **left cosets** of $\mathfrak{G}$ mod $\mathfrak{H}$. That is, for fixed α in $\mathfrak{G}$ the set of elements $\alpha\mathfrak{H} = \{\alpha\xi\}$, ξ in $\mathfrak{H}$, is a left coset. It is easy to see that the collection of left cosets $\{\alpha\mathfrak{H}\}$ is a covariant decomposition into layers. Consequently, the subgroups $\mathfrak{H}$ of $\mathfrak{G}$ and the covariant decompositions of $\mathfrak{G}$ (under left translations) are in one-one correspondence. A second decomposition is obtained by considering right cosets. If these decompositions coincide, the subgroup $\mathfrak{H}$ is called **normal** or **invariant.**

2. Similarity

Let $\mathbf{S}_1 = \{p_1, q_1, \cdots\}$ be a set of points and $\mathfrak{G}_1 = \{\alpha_1, \beta_1, \gamma_1, \cdots\}$ a group of transformations of $\mathbf{S}_1$ onto $\mathbf{S}_1$. Let ω be a one-one mapping between $\mathbf{S}_1$ and a space $\mathbf{S}_2 = \{p_2, q_2, \cdots\}$. We define a group $\mathfrak{G}_2 = \{\alpha_2, \beta_2, \cdots\}$ on $\mathbf{S}_2$ by the following process: If $\alpha_1 p_1 = q_1$, $\omega p_1 = p_2$, and $\omega q_1 = q_2$, then the mapping α_2: $q_2 = \omega\alpha_1\omega^{-1}p_2$ is a transformation of $\mathbf{S}_2$ onto $\mathbf{S}_2$. It is clear that these transformations form a group isomorphic to the group $\mathfrak{G}_1$. This group $\mathfrak{G}_2$ is said to be **similar** to $\mathfrak{G}_1$.

By definition, similar transformation groups are isomorphic, but in general isomorphic transformation groups are not necessarily similar. To see this, we consider the group $\mathfrak{G}_1$ consisting of the translations on a line

$$x' = x + a,$$

and the group $\mathfrak{G}_2$ consisting of the special class of translations in the plane

$$\begin{cases} x' = x + a \\ y' = y. \end{cases}$$

These groups are clearly isomorphic, but $\mathfrak{G}_1$ is transitive while $\mathfrak{G}_2$ is intransitive, since two points in the plane can be mapped into each other only if they have the same y-coordinate. The fact that $\mathfrak{G}_1$ and $\mathfrak{G}_2$ are not similar follows from the next lemma.

LEMMA 2.1. *Let $\mathfrak{G}_1$ and $\mathfrak{G}_2$ be similar groups and assume $\mathfrak{G}_1$ is transitive. Then $\mathfrak{G}_2$ is transitive.*

Proof. Let p_2, q_2 be arbitrary points in $\mathbf{S}_2$. Let ω be the given one-one mapping of $\mathbf{S}_1$ onto $\mathbf{S}_2$. Define $p_1 = \omega^{-1}p_2$ and $q_1 = \omega^{-1}q_2$. Since $\mathfrak{G}_1$ is transitive, there is an α in $\mathfrak{G}_1$ such that $\alpha p_1 = q_1$. Then $q_2 = \omega\alpha\omega^{-1}p_2$, so $\omega\alpha\omega^{-1}$ is an element of $\mathfrak{G}_2$ which takes p_2 into q_2. Hence $\mathfrak{G}_2$ is transitive.

If in the last example we consider continuous transformations, then the fact that the line and plane are of different dimensions establishes the impossibility of similarity between the groups. However, we note that the result was obtained using only the one-one character of the mapping ω.

If the spaces $\mathbf{S}_1$ and $\mathbf{S}_2$ are the same, then the mapping ω is a mapping of $\mathbf{S}$ onto itself. In this case the group $\mathfrak{G}_2$ generated from $\mathfrak{G}_1$ by transformations of the form $\omega\alpha\omega^{-1}$ (α in $\mathfrak{G}_1$) is again a group of transformations of $\mathbf{S}$ onto $\mathbf{S}$. In general this group is not the same as the original group $\mathfrak{G}_1$. However, if we take for ω an element α of $\mathfrak{G}_1$, then it is obvious that $\mathfrak{G}_2$ will coincide with $\mathfrak{G}_1$. It is possible to take for ω a mapping of $\mathbf{S}$ onto $\mathbf{S}$ which is not in $\mathfrak{G}_1$ and still have $\mathfrak{G}_2$ coincide with $\mathfrak{G}_1$. As an example, we take $\mathbf{S}$ to be the Euclidean three space and $\mathfrak{G}_1$ to be the orthogonal group (see Example 6, above). For ω we select

$$\omega : \begin{cases} x' = ax \\ y' = ay \\ z' = az \end{cases} \qquad a \neq 0.$$

Then any transformation $\omega\alpha\omega^{-1}$ dilates an element by the distance a^{-1}, transforms it by α, and then contracts it by an amount a. Hence distances are preserved, and $\omega\alpha\omega^{-1}$ is an element in $\mathfrak{G}_1$, even though ω obviously is not if $a \neq 1$.

On the other hand, in the hyperbolic geometry of Bolyai-Lobachevski, the only transformations ω that send each motion α into a motion $\omega\alpha\omega^{-1}$ are the motions themselves.

In general the transformations ω of **S** such that $\omega\mathfrak{G}\omega^{-1} = \mathfrak{G}$ form a group (of which $\mathfrak{G}$ is a subgroup) called the **similarity group** of $\mathfrak{G}$.

LECTURE II

3. Representations of Groups

Let 𝔊 be a group of transformations on a set **S**. As we have already seen, 𝔊 generates other groups of transformations acting on various types of figures in **S**, for example, on a covariant layer decomposition. Thus we may have another space **T** on which each α in 𝔊 induces a transformation τ_α and such that

$$\tau_{\alpha\beta} = \tau_\alpha \tau_\beta.$$

As an example, **T** could be the space of pairs of elements of **S**. In many cases, the mapping $\alpha \longrightarrow \tau_\alpha$ is an isomorphism on 𝔊 onto a transformation group of **T**, but in general it is only a homomorphism. It is clear that in describing this mapping, the fact that the group 𝔊 is a transformation group acting on **S** is quite immaterial; what is essential is the abstract group 𝔤 associated with 𝔊.

In general we use this notation: lower case Gothic 𝔤 for an abstract group, capital Gothic 𝔊 for a transformation group that is a **realization** of the abstract group 𝔤.

In many problems of analysis and geometry, it is essential to find all realizations of a given transformation group. The problem of doing so originated in the theory of algebraic invariants, which itself stems from projective geometry.

Let 𝔊 be the group of all linear homogeneous transformations in n-space:

$$\mathbf{y} = A\mathbf{x},$$

where

$$\mathbf{x} = \begin{bmatrix} x_1 \\ \cdot \\ \cdot \\ \cdot \\ x_n \end{bmatrix}, \quad \mathbf{y} = \begin{bmatrix} y_1 \\ \cdot \\ \cdot \\ \cdot \\ y_n \end{bmatrix}, \quad A = \begin{bmatrix} a_{11} \cdots a_{1n} \\ \cdot \\ \cdot \\ \cdot \\ a_{n1} \cdots a_{nn} \end{bmatrix}, \quad \det(A) \neq 0.$$

We consider homogeneous equations in the x_i as follows:

$$\textbf{hyperplane:} \qquad \sum_{i=1}^{n} a_i x_i = 0,$$

$$\textbf{quadric:} \qquad \sum_{i,j=1}^{n} a_{ij} x_i x_j = 0,$$

etc. Each element of $\mathfrak{G}$ subjects the coefficients of these forms to linear trans-formations. Thus we obtain realizations of $\mathfrak{G}$, one on the space of linear forms, another on the space of quadratic forms, etc.

We obtain further examples in differential geometry. Here we consider a group $\mathfrak{G}_1$ of transformations of the form

$$y_i = f_i(x_1, \cdots, x_n), \qquad i = 1, \cdots, n,$$

where suitable differentiability conditions are imposed on the functions f_i. We have

$$dy_i = \sum_{\rho=1}^{n} \frac{\partial f}{\partial x_\rho} dx_\rho,$$

which means that each transformation of the group induces a linear transforma-tion (at each point) on the space of differentials. A Riemannian metric is given by a positive definite quadratic form

$$\sum_{i,j=1}^{n} g_{ij}(x) \, dx_i \, dx_j.$$

Each transformation of the group $\mathfrak{G}_1$ induces a further linear trans-formation on the coefficients of this form. Thus we have another realization of $\mathfrak{G}_1$. Similar remarks apply to cubic forms, etc.

We shall be concerned mainly with realizations by means of linear transformations. We pass to some basic definitions:

Let $\mathfrak{g}$ be an abstract group. By a **quantity** (German: **Grösse**) is meant an element of a set **S** on which a realization of $\mathfrak{g}$ acts.

A realization $\alpha \longrightarrow \tau_\alpha$ of $\mathfrak{g}$ is called **faithful** provided it is an isomorph-ism, i.e., if $\alpha \neq \beta$, then $\tau_\alpha \neq \tau_\beta$.

A **representation** of $\mathfrak{g}$ is a realization by means of linear transformations. These definitions combined yield that of a **faithful representation.**

In the middle of the last century, A. Cayley started the theory of group representations in connection with invariant theory. Later, G. Frobenius and W. Burnside did major work on the subject.

4. Combinations of Representations

Let $\alpha \longrightarrow U(\alpha)$ be a representation of a group $\mathfrak{g}$. Here $U(\alpha) = \|u_{ij}(\alpha)\|$ is an $n \times n$ non-singular matrix associated with the group element α. The integer n is called the **degree** of the representation. The associated linear transformation is given by

$$\mathbf{x} \longrightarrow \mathbf{y} = U(\alpha)\mathbf{x}.$$

We shall study several important methods of generating new representations from a given pair of representations. But first we consider these methods as applied to individual linear transformations.

Suppose U_1, U_2 are matrices of degrees n_1 and n_2 that act on vectors denoted $\mathbf{x}^1$ and $\mathbf{x}^2$ respectively. We form $\mathbf{x} = (\mathbf{x}^1, \mathbf{x}^2)$, with the operations of addition and scalar multiplication defined componentwise. The transformation U defined by

$$U\mathbf{x} = (U_1\mathbf{x}^1, U_2\mathbf{x}^2)$$

evidently has the matrix

$$\begin{pmatrix} U_1 & 0 \\ 0 & U_2 \end{pmatrix}$$

of degree $n_1 + n_2$. We call this the result of **addition** of the transformations and write it $U_1 + U_2$. In particular, if $U_1(\alpha)$ and $U_2(\alpha)$ are representations of the group $\mathfrak{g}$, then

$$U(\alpha) = \begin{pmatrix} U_1(\alpha) & 0 \\ 0 & U_2(\alpha) \end{pmatrix}$$

is a representation of $\mathfrak{g}$, the result of the **addition** of the given representations $U_1(\alpha)$ and $U_2(\alpha)$.

Another possibility for combining representations is **multiplication.** From the vectors

$$\mathbf{x}^1 = (x_1^1, \cdots, x_{n_1}^1)' \quad \text{and} \quad \mathbf{x}^2 = (x_1^2, \cdots, x_{n_2}^2)',$$

we form the $n_1 n_2$ products

$$x_{ij} = x_i^1 x_j^2.$$

We observe that these products transform in a definite way when the vectors $\mathbf{x}^1$ and $\mathbf{x}^2$ are transformed according to the rules

$$\mathbf{y}^1 = U_1\mathbf{x}^1, \qquad \mathbf{y}^2 = U_2\mathbf{x}^2.$$

We have

$$y_i^1 y_j^2 = \sum_{\rho,\sigma} u_{i\rho}^1 u_{j\sigma}^2 x_\rho^1 x_\sigma^2, \qquad \begin{cases} 1 \le \rho \le n_1 \\ 1 \le \sigma \le n_2, \end{cases}$$

which we may write as

$$y_i^1 y_j^2 = \sum_{\rho,\sigma} b_{ij\rho\sigma} x_\rho^1 x_\sigma^2 \qquad \text{with} \qquad b_{ij\rho\sigma} = u_{i\rho}^1 u_{j\sigma}^2.$$

The scalars $b_{ij\rho\sigma}$ are uniquely determined. In fact, if also

$$y_i^1 y_j^2 = \sum_{\rho,\sigma} c_{ij\rho\sigma} x_\rho^1 x_\sigma^2,$$

then

$$(b_{ij\rho\sigma} - c_{ij\rho\sigma}) x_\rho^1 x_\sigma^2 = 0.$$

This implies

$$b_{ij\rho\sigma} - c_{ij\rho\sigma} = 0,$$

since *the $n_1 n_2$ products $x_\rho^1 x_\sigma^2$ are linearly independent*. Thus we define a transformation, denoted $U = U_1 \times U_2$, on the space of vectors (x_{ij}). This result of **multiplication** of U_1 and U_2 is of degree $n_1 n_2$ and is given by

$$y_{ij} = \sum_{\rho,\sigma} b_{ij\rho\sigma} x_{\rho\sigma}.$$

(The concept of multiplication of representations was introduced by A. Hurwitz about 1890.) With this idea, we may introduce $U \times U$, $U_1 \times U_2 \times U_3$, etc. If $U_1(\alpha)$ and $U_2(\alpha)$ are representations of a group, then so is $U_1(\alpha) \times U_2(\alpha)$.

It is worth noting that the space of products $x_i^1 x_j^2$ decomposes into co-variant layers according to equations

$$x_i^1 x_j^2 = c_{ij} = \text{constants, not all zero.}$$

The only transformations not changing the layers are those of the form

$$x_i^1 \longrightarrow c x_i^1, \quad x_j^2 \longrightarrow c^{-1} x_j^2 \qquad \text{where} \quad c \neq 0.$$

If in the case above $U_1 = U_2 = U$, and if we also identify $\mathbf{x}^1 = \mathbf{x}^2 = \mathbf{x}$, we obtain the **square** of U:

$$y_i y_j = \sum_{\rho,\sigma} u_{i\rho} u_{j\sigma} x_\rho x_\sigma.$$

Since there are $n(n+1)/2$ linearly independent components $x_i x_j$, we obtain a transformation of degree $n(n+1)/2$. Similarly, we have the **cube** of U,

$$y_i y_j y_k = \sum_{\rho,\sigma,\tau} u_{i\rho} u_{j\sigma} u_{k\tau} x_\rho x_\sigma x_\tau, \text{ etc.,}$$

leading to higher **powers** of U. Applied to a representation $U(\alpha)$, this process yields **powers** of the representation.

Yet another method of generating new representations is obtained by the formation of the **Grassmann quantities.** Given vectors $\mathbf{x}^1 = (x_1^1, \cdots, x_n)$, $\mathbf{x}^2 = (x_1^2, \cdots, x_n^2)$ in the same space, we form all 2×2 determinants

$$\begin{vmatrix} x_i^1 & x_j^1 \\ x_i^2 & x_j^2 \end{vmatrix}, \qquad i < j.$$

These are the components of the Grassmann quantities. If $\mathbf{y} = U\mathbf{x}$ is a transformation, then an elementary rule for expanding determinants yields

$$\begin{vmatrix} y_i^1 & y_j^1 \\ y_i^2 & y_j^2 \end{vmatrix} = \sum_{\rho<\sigma} \begin{vmatrix} u_{i\rho} & u_{i\sigma} \\ u_{j\rho} & u_{j\sigma} \end{vmatrix} \cdot \begin{vmatrix} x_\rho^1 & x_\sigma^1 \\ x_\rho^2 & x_\sigma^2 \end{vmatrix}, \qquad (i < j).$$

The determinants $|x_\alpha{}^k|$ are linearly independent functions; hence we have uniquely determined a linear transformation

$$y_{ij} = \sum_{\rho<\sigma} \begin{vmatrix} u_{i\rho} & u_{i\sigma} \\ u_{j\rho} & u_{j\sigma} \end{vmatrix} x_{\rho\sigma} \qquad (i < j)$$

of degree $\binom{n}{2} = n(n - 1)/2$. It is often convenient to let the indices i, j in x_{ij} range independently from 1 to n and make the restriction $x_{ji} = -x_{ij}, x_{ii} = 0$ so that we have in effect the space of antisymmetric tensors. In case $U(\alpha)$ is a representation of degree n, the induced transformations on the Grassmann quantities form a representation of degree $n(n - 1)/2$. Similarly we may form Grassmann quantities of order 3 using the 3×3 determinants formed by three vectors, etc.

Finally, we note that a transformation U on a space of vectors $\mathbf{x}$ induces a transformation on the space of linear forms $\mathbf{l}(\mathbf{x}) = \sum l_i x_i$ on the space. In fact, if $\mathbf{y} = U\mathbf{x}$, then $\mathbf{l}(\mathbf{x}) = \sum m_i y_i$ and we have

$$\mathbf{y} = U\mathbf{x}, \qquad \mathbf{m} = (U^{-1})'\mathbf{l} = (U')^{-1}\mathbf{l}.$$

We say that $\mathbf{x}$ and $\mathbf{l}$ are transformed **contragrediently.**

5. Similarity and Reducibility

Let $U(\alpha)$ be a representation of degree n, and let C be a non-singular $n \times n$ matrix. Then

$$V(\alpha) = C\,U(\alpha)\,C^{-1}$$

is another representation, and we say that $U(\alpha)$ and $V(\alpha)$ are **similar representations.** We may consider C either as a mapping to a new space or as a transition of coordinates. If $\mathbf{x}' = C\mathbf{x}$ is the associated transformation, then from $\mathbf{y}' = C\mathbf{y}$ and $\mathbf{y} = U(\alpha)\mathbf{x}$ we have

$$\mathbf{y}' = C\,U(\alpha)\,\mathbf{x} = C\,U(\alpha)\,C^{-1}\mathbf{x}' = V(\alpha)\,\mathbf{x}'.$$

We shall not distinguish between similar representations, since a mere change of coordinates brings one to the other.

DEFINITION 5.1. A representation of degree n is called **reducible** if there is a non-zero subspace of lower dimension which is invariant under all transformations $U(\alpha)$ of the representation. If not reducible, the representation is called **irreducible.**

Note. This definition may be applied to any system of linear transformations, whether or not it is a group.

Example. Consider the group

$$\begin{cases} x' = ax \\ y' = bx + cy \end{cases} \qquad a \neq 0, \quad c \neq 0.$$

Then $x = 0$ defines an invariant one-dimensional subspace, hence the group is reducible.

Example. Consider the orthogonal group in the plane:

$$\begin{cases} x' = a_1 x + a_2 y \\ y' = -a_2 x + a_1 y \end{cases} \qquad a_1{}^2 + a_2{}^2 = 1.$$

If the a_i are real, this group is irreducible, since any line may be rotated to a new position. But if the a_i are allowed to be complex, the lines $y = ix$ and $y = -ix$ are invariant one-dimensional subspaces. Proof for the case $y = ix$:

$$x' = a_1 x + a_2 y = (a_1 + ia_2)x, \qquad y' = -a_2 x + a_1 y = i(ia_2 + a_1)x,$$

hence $y' = ix'$.

Now suppose $U(\alpha)$ is a representation by means of transformations on the vector space **L** and that **M** is a subspace reduced by $U(\alpha)$, i.e., invariant under $U(\alpha)$. We observe that the linear varieties in **L** parallel to **M** are transformed amongst themselves by the $U(\alpha)$; hence we have a covariant decomposition of **L** into layers. In this situation, we associate two new representations of the group:

1. layers $\longrightarrow$ layers,

2. **M** $\longrightarrow$ **M**.

We suppose dim (**L**) $= n$ and dim (**M**) $= m$ with $0 < m < n$. We select a coordinate system of **L**, so that **M** is given by the relations

$$x_1 = x_2 = \cdots = x_{n-m} = 0.$$

If a transformation $\mathbf{y} = U\mathbf{x}$ leaves $\mathbf{M}$ invariant, then we must have

$$y_1 = y_2 = \cdots = y_{n-m} = 0$$

if $\mathbf{x}$ is in $\mathbf{M}$, hence the transformation U has, in these coordinates, the form

($\dagger$)
$$\begin{cases}
y_1 = u_{1,1}x_1 + \cdots + u_{1,\,n-m}x_{n-m} \\
\quad \cdot \quad \cdot \quad \cdot \quad \cdot \quad \cdot \quad \cdot \quad \cdot \quad \cdot \quad \cdot \\
y_{n-m} = u_{n-m,1}x_1 + \cdots + u_{n-m,\,n-m}x_{n-m} \\
y_{n-m+1} = u_{n-m+1,1}x_1 + \cdots + u_{n-m+1,\,n}x_n \\
\quad \cdot \quad \cdot \quad \cdot \quad \cdot \quad \cdot \quad \cdot \quad \cdot \quad \cdot \quad \cdot \\
y_n = u_{n,1}x_1 + \cdots + u_{n,\,n}x_n,
\end{cases}$$

which means that the corresponding matrix has the form

$$U = \begin{pmatrix} U' & 0 \\ * & U'' \end{pmatrix}, \qquad \begin{array}{l} U' \text{ of degree } (n-m) \\ U'' \text{ of degree } m. \end{array}$$

A linear variety parallel to $\mathbf{M}$ (i.e., a layer) is determined by equations of the form $x_1 = c_1, \cdots, x_{n-m} = c_{n-m}$ with the c_i constant. This variety goes into the layer given by the relations

$$y_i = \sum_{j=1}^{n-m} u_{ij}c_j \qquad (i = 1, \cdots, n-m).$$

Thus we see that U' characterizes the transformation induced on the layer space by U.

On the other hand, a point of $\mathbf{M}$ is characterized by the values of $x_{n-m+1}, \cdots, x_n$. The image under U of such a point is evidently given by the last m lines of ($\dagger$) with $x_1 = \cdots = x_{n-m} = 0$; hence the transformation U induced on $\mathbf{M}$ is given by U''.

We have arrived at the following result: If a subspace $\mathbf{M}$ reduces a linear transformation U, then there exists a non-singular matrix C (i.e., a transition of coordinates to a coordinate system fitted to $\mathbf{M}$) such that

$$C U C^{-1} = \begin{pmatrix} U' & 0 \\ * & U'' \end{pmatrix}.$$

If $U(\alpha)$ is a representation reduced by $\mathbf{M}$, then there is a non-singular constant matrix C such that

$$C U(\alpha) C^{-1} = \begin{pmatrix} U'(\alpha) & 0 \\ * & U''(\alpha) \end{pmatrix}.$$

for each α. We easily see that

$$U'(\alpha\beta) = U'(\alpha)U'(\beta), \qquad U''(\alpha\beta) = U''(\alpha)U''(\beta),$$

so that $U'(\alpha)$ and $U''(\alpha)$ are new representations of lower degree than U.

DEFINITION 5.2. A set of linear transformations on **L** is called **completely reducible** provided that whenever **M** is a proper subspace of **L** reduced by the set, then there is a supplementary subspace **N** also reduced by the set. By **supplementary,** we mean that **L** is spanned by **M** and **N**, **L** = **M** + **N**, and dim (**L**) = dim (**M**) + dim (**N**). (Consequently **M** ∩ **N** = **0**.)

In this situation, a coordinate system may be adapted to **M** and **N** so that **M** is given by $x_1 = \cdots = x_{n-m} = 0$ and **N** by $x_{n-m+1} = \cdots = x_n = 0$. With respect to this coordinate system, each transformation of the set has the form

$$U = \begin{pmatrix} U' & 0 \\ 0 & U'' \end{pmatrix}.$$

If the matrix of U were given instead in some other coordinate system, the transition of coordinates would yield a non-singular matrix C such that

$$C U C^{-1} = \begin{pmatrix} U' & 0 \\ 0 & U'' \end{pmatrix}.$$

Example. Consider the group

$$\begin{cases} x' = ax \\ y' = bx + cy \end{cases} \qquad a \neq 0, \quad c \neq 0.$$

Here the subspace **M** given by $x = 0$ is invariant, but there is no other invariant line. Hence the group fails to be completely reducible.

We now consider some important cases of completely reducible groups.

THEOREM 5.3. *If $\mathfrak{G}$ is a group of orthogonal real linear transformations on a vector space* **L**, *then $\mathfrak{G}$ is completely reducible.*

Proof. Suppose there is a subspace **M** of **L** which reduces $\mathfrak{G}$. The subspace **N** of **L** orthogonal to **M** also reduces $\mathfrak{G}$ since orthogonal vectors remain orthogonal under orthogonal transformations. But **M** and **N** are supplementary, hence $\mathfrak{G}$ is completely reducible.

THEOREM 5.4. *Any finite group $\mathfrak{G}$ of real linear transformations on a vector space* **L** *is similar to a group of orthogonal transformations.*

Proof. Let $F = \sum a_{\rho\sigma} x_\rho x_\sigma$ be a positive definite quadratic form on **L**; here $a_{\sigma\rho} = a_{\rho\sigma}$. We suppose $\mathfrak{G} = \{U_1, \cdots, U_g\}$. Each U_i transforms F to a new positive definite quadratic form F_i; we form the average

$$E = \frac{1}{g} \sum_{i=1}^{g} F_i,$$

also a positive definite quadratic form. But E is invariant under each U in $\mathfrak{G}$. For if we denote the F_i by F^{U_i}, $i = 1, 2, \cdots, g$, and the transform of E under an arbitrary transformation U of $\mathfrak{G}$ by E^U, then

$$E^U = \frac{1}{g}\left(\sum_{i=1}^{g} F^{UU_i}\right) = \frac{1}{g}\sum_{j=1}^{g} F^{U_j} = E,$$

since the mapping $U_i \longrightarrow UU_i$ merely effects a permutation of the elements of the finite group $\mathfrak{G}$. Thus we have found a positive definite quadratic form E that is invariant under $\mathfrak{G}$. In a suitably chosen coordinate system, E is the unit form

$$E = \sum x_i^2,$$

and the transformations in $\mathfrak{G}$ are orthogonal.

We have the following analogous results in the complex case.

THEOREM 5.5. *Let $\mathfrak{G}$ be a group of unitary linear transformations on a vector space $\mathbf{L}$ over the field of complex numbers. Then $\mathfrak{G}$ is completely reducible.*

THEOREM 5.6. *Let $\mathfrak{G}$ be any finite group of complex linear transformations. Then $\mathfrak{G}$ is similar to a group of unitary transformations.*

We recall that we must deal here with positive definite Hermitian forms $H = \sum a_{\rho\sigma}x_\rho\bar{x}_\sigma$, where $a_{\sigma\rho} = \bar{a}_{\rho\sigma}$, that the unit form is $I = \sum x_i\bar{x}_i$, and that the orthogonality of two vectors $\mathbf{x}$ and $\mathbf{y}$ is given by $\sum x_i\bar{y}_i = 0$. With this in mind, we see that the proofs in the complex case are virtually identical with those in the real case.

THEOREM 5.7. *Let $\mathfrak{G}$ be a finite group of real or complex linear transformations. Then $\mathfrak{G}$ is completely reducible.*

This is an immediate consequence of the preceding results.

DEFINITION 5.8. A set of linear transformations on a vector space $\mathbf{L}$ is said to be **completely decomposed** if there are linearly independent subspaces $\mathbf{M}_1, \cdots, \mathbf{M}_k$ of $\mathbf{L}$ which together span $\mathbf{L}$, and such that each is invariant and irreducible.

Evidently in this situation each matrix U of the set is a matrix of diagonal blocks,

$$U = \operatorname{diag}\{U', U'', \cdots, U'''\},$$

(zeros elsewhere) with respect to a suitable coordinate system. In case of a representation, each $U'(\alpha)$, $U''(\alpha)$, etc. is an irreducible representation. This decomposition may be achieved for any completely reducible representation.

Example. We consider the product representation $U(\alpha) \times U(\alpha)$ of the group of all homogeneous linear transformations:

$$y_{ij} = \sum_{\rho,\sigma} u_{i\rho}(\alpha) u_{j\sigma}(\alpha) x_{\rho\sigma}.$$

The subspace **M** of symmetric tensors given by $x_{\rho\sigma} = x_{\sigma\rho}$ is invariant, as is the subspace **N** of antisymmetric tensors given by $x_{\rho\sigma} = -x_{\sigma\rho}$. Here

$$\dim(\textbf{M}) + \dim(\textbf{N}) = \frac{n(n+1)}{2} + \frac{n(n-1)}{2} = n^2 = \dim(\textbf{L}).$$

Also **M** and **N** span **L**, since each vector $x_{\rho\sigma}$ has the decomposition

$$x_{\rho\sigma} = \frac{x_{\rho\sigma} + x_{\sigma\rho}}{2} + \frac{x_{\rho\sigma} - x_{\sigma\rho}}{2}.$$

We note without proof the fact that both **M** and **N** are *irreducible* so that the tensor space **L** is completely decomposed into **M** and **N**.

LECTURE III

6. Representations of Cyclic Groups

Let $\mathfrak{g}$ be a cyclic group, i.e., $\mathfrak{g} = \{\epsilon = \alpha^0, \alpha^1, \cdots, \alpha^{g-1}\}$, $\alpha^g = \epsilon$. We shall determine the totality of irreducible representations of $\mathfrak{g}$ over the field of complex numbers. Given a representation in n-dimensional complex space, we let $\alpha \longrightarrow U$. Since $\mathfrak{g}$ is cyclic, we have the correspondences $\alpha^i \longrightarrow U^i$, $i = 1, 2, \cdots, g$, and therefore we find $U^g = I$, where I is the unit matrix. From elementary matrix theory it is clear that if a power of a matrix U is the identity, then there is a coordinate system in which U has diagonal form

$$U = \text{diag } \{\lambda_1, \cdots, \lambda_n\}.$$

In this coordinate system we have the correspondence

$$\alpha^k \longrightarrow U^k = \text{diag } \{\lambda_1{}^k, \cdots, \lambda_n{}^k\}, \qquad k = 1, \cdots, g.$$

The mapping $\alpha^k \longrightarrow \lambda_j{}^k$, is itself a one-dimensional representation of $\mathfrak{g}$ for each $j = 1, 2, \cdots, n$. Hence we have the following theorem.

THEOREM 6.1. *Any representation of a cyclic group $\mathfrak{g}$ can be decomposed into one-dimensional representations.*

From the relation $U^g = I$ we see that $\lambda_j{}^g = 1$ for $j = 1, 2, \cdots, n$, so each λ_j must be a gth root of unity.

THEOREM 6.2. *The distinct one-dimensional representations of a cyclic group $\mathfrak{g}$ are the representations $\alpha^k \longrightarrow \lambda^k$, where λ is any one of the gth roots of unity.*

7. Representations of Finite Abelian Groups

Let $\mathfrak{g}$ be abelian and $\alpha \longrightarrow U(\alpha)$ an n-dimensional representation. The relation $\alpha\beta = \beta\alpha$ for α, β in $\mathfrak{g}$ implies that $U(\alpha)\,U(\beta) = U(\beta)\,U(\alpha)$; hence the matrices $U(\alpha)$ form a commutative system.

LEMMA 7.1. *If a system $\mathbf{\Sigma}$ of linear transformations U is abelian and each matrix U of $\mathbf{\Sigma}$ can be diagonalized, then there is a coordinate system in which all matrices of $\mathbf{\Sigma}$ are diagonal.*

Proof. Consider a coordinate system in which one of the elements of $\mathbf{\Sigma}$, say U_1, is diagonal:

$$U_1 = \operatorname{diag}\{\lambda_1, \cdots, \lambda_n\},$$

with not all λ_i equal. Suppose there is a block of k' values equal to λ', k'' equal to λ'', etc. That is,

$$U_i = \operatorname{diag}\{\lambda' I_{k'}, \lambda'' I_{k''}, \cdots\}.$$

Let U be a matrix of $\mathbf{\Sigma}$, which we decompose into blocks:

$$U = \begin{bmatrix} U_{11} & U_{12} & \cdots \\ U_{21} & U_{22} & \cdots \\ \cdot & \cdot & \\ \cdot & \cdot & \\ \cdot & \cdot & \end{bmatrix},$$

where U_{11} is a $k' \times k'$ matrix, U_{12} is a $k' \times k''$ matrix, U_{22} is a $k'' \times k''$ matrix, etc.

From the relation $U_1 U = U U_1$, we have

$$U_1 U = \begin{bmatrix} \lambda' U_{11} & \lambda' U_{12} & \cdots \\ \lambda'' U_{21} & \lambda'' U_{22} & \cdots \\ \cdot & \cdot & \\ \cdot & \cdot & \\ \cdot & \cdot & \end{bmatrix} = \begin{bmatrix} U_{11}\lambda' & U_{12}\lambda'' & \cdots \\ U_{21}\lambda' & U_{22}\lambda'' & \cdots \\ \cdot & \cdot & \\ \cdot & \cdot & \\ \cdot & \cdot & \end{bmatrix} = U U_1.$$

This implies that $\lambda' U_{12} = \lambda'' U_{12}$, $\lambda'' U_{21} = \lambda' U_{21}$, $\cdots$, and since $\lambda' \neq \lambda''$ by hypothesis, we find that $U_{12} = U_{21} = 0$. Hence the matrix U must have the form

$$U = \operatorname{diag}\{U_{11}, U_{22}, \cdots\}.$$

However, the same argument applies to each of the matrices U_{ii}, since we know from the elementary theory of matrices that the possibility of the diagonalization of U implies that each U_{ii} can be diagonalized. By induction we conclude that U must be a diagonal matrix.

THEOREM 7.2. *Let $\mathfrak{g}$ be a finite abelian group and let* $\alpha \longrightarrow U(\alpha)$ *be a representation of* $\mathfrak{g}$. *Then the matrices* $U(\alpha)$ *are simultaneously diagonalizable. That is, the representation can be decomposed into the sum of one-dimensional representations.*

Proof. It is clear that the system $\{U(\alpha)\}$ is a commutative one. For each α in $\mathfrak{g}$ we have $\alpha^g = \epsilon$, where g is the order of $\mathfrak{g}$. Hence $U^g = I$. Since a power of U is the identity, U may be transformed to diagonal form. Applying Lemma 7.1, we obtain the result.

We now know that each representation of a finite abelian group $\mathfrak{g}$ can be decomposed into one-dimensional representations; thus we seek all possible one-dimensional representations of $\mathfrak{g}$. For this purpose we employ the following standard decomposition theorem.

THEOREM 7.3. *Each finite abelian group* $\mathfrak{g}$ *is the direct product*

$$\mathfrak{g} = \mathfrak{g}_1 \times \mathfrak{g}_2 \times \mathfrak{g}_3 \times \cdots \times \mathfrak{g}_p,$$

where $\mathfrak{g}_i$ *is a cyclic group of order* g_i *with generator* α_i, $i = 1, 2, \cdots, p$. *Each element* α *of* $\mathfrak{g}$ *is uniquely representable in the form*

$$\alpha = \alpha_1{}^{n_1} \alpha_2{}^{n_2} \cdots \alpha_p{}^{n_p}, \qquad 0 \le n_i \le g_i - 1, \quad i = 1, 2, \cdots, p.$$

Let $\xi \longrightarrow \lambda(\xi)$ be a one-dimensional representation, and let ξ be in $\mathfrak{g}$. Then from the factorization

$$\xi = \alpha_1{}^{n_1} \alpha_2{}^{n_2} \cdots \alpha_p{}^{n_p},$$

we have the relation

$$\lambda(\xi) = [\lambda(\alpha_1)]^{n_1} [\lambda(\alpha_2)]^{n_2} \cdots [\lambda(\alpha_p)]^{n_p}.$$

Each of the quantities $\lambda_1 = \lambda(\alpha_1), \cdots, \lambda_p = \lambda(\alpha_p)$ is a $g_1{}^{\text{th}}, \cdots, g_p{}^{\text{th}}$ root of unity respectively. Again we see immediately that by substituting for the λ_i any such roots of unity into the formula for $\lambda(\xi)$, we obtain a representation of $\mathfrak{g}$. The *number* of one-dimensional representations is therefore $g_1 g_2 \cdots g_p = g$.

Let $\mathfrak{g}$ be a group of non-singular matrices $\{U\}$. Each matrix of $\mathfrak{g}$ may be considered as a point in n^2-dimensional complex Euclidean space, the n^2 elements of the matrix representing the coordinates of this point. The group $\mathfrak{g}$ is called **compact** if the matrices of $\mathfrak{g}$, considered as points in the n^2-dimensional Euclidean space, form a closed and bounded set.

THEOREM 7.4. *Let* $\mathfrak{g}$ *be a compact abelian group of non-singular matrices. Then the matrices of* $\mathfrak{g}$ *can be simultaneously diagonalized.*

Proof. Let A be an element of $\mathfrak{g}$. We shall show that A can be diagonalized. Form the matrices A^m, $m = \pm 1$, ± 2, $\cdots$. These are in $\mathfrak{g}$, and hence the elements of A^m are bounded for all m. If λ', λ'', $\cdots$ are the distinct characteristic values of A, then, in a suitable coordinate system, A is the sum of diagonal blocks of the form

$$
\begin{bmatrix}
\lambda' & 0 & \cdot & \cdot & 0 \\
 & \cdot & & & \cdot \\
 & & \cdot & & \cdot \\
 & * & & \cdot & 0 \\
 & & & & \lambda'
\end{bmatrix},
$$

where the * indicates that any values may occur in those positions of the matrix. Consequently, the matrix A^m is the sum of diagonal blocks

$$
\begin{bmatrix}
(\lambda')^m & 0 & \cdot & \cdot & 0 \\
 & \cdot & & & \cdot \\
 & & \cdot & & \cdot \\
 & * & & \cdot & 0 \\
 & & & & (\lambda')^m
\end{bmatrix}.
$$

From the compactness, we conclude that $|\lambda'|^m \le M$ for all positive and negative integers m. Hence $|\lambda'| = 1$. Since the same argument applies to the remaining characteristic values, we have $|\lambda'| = |\lambda''| = |\lambda'''| = \cdots = 1$.

For simplicity we first suppose $\lambda' = 1$ and denote by A_1 the first block in the representation of A,

$$
A_1 = \begin{bmatrix}
1 & 0 & \cdot & \cdot & 0 \\
 & \cdot & & & \cdot \\
 & & \cdot & & \cdot \\
 & * & & \cdot & 0 \\
 & & & & 1
\end{bmatrix}.
$$

We write A_1 in the form $A_1 = I + F$, where I is the identity and

$$
F = \begin{bmatrix}
0 & 0 & \cdot & \cdot & 0 \\
 & \cdot & & & \cdot \\
 & & \cdot & & \cdot \\
 & * & & \cdot & 0 \\
 & & & & 0
\end{bmatrix}.
$$

We form the powers

$$A_1^m = (I + F)^m = I + \binom{m}{1}F + \binom{m}{2}F^2 + \cdots.$$

A simple computation shows that the quantity F^2 has the form

$$F^2 = \begin{bmatrix} 0 & \cdot & \cdot & \cdot & \cdot & 0 \\ 0 & \cdot & & & & \cdot \\ & 0 & \cdot & & & \cdot \\ & & \cdot & \cdot & & \cdot \\ & * & & \cdot & \cdot & \cdot \\ & & & & 0 & 0 \end{bmatrix},$$

that is, F^2 has zeros on and above the main diagonal and on the first sub-diagonal. The matrix F^3 has zeros at all the places where F^2 is zero and in addition at all elements of the second sub-diagonal, and so forth. The elements of $(I + F)^m$ which are part of A^m remain bounded. However, in the expansion of $(I + F)^m$ the elements of the first sub-diagonal receive contributions from $\binom{m}{1}F$ only, the remaining powers of F all being zero along this sub-diagonal. Allowing m to approach infinity, we conclude that the elements along this sub-diagonal are zero. Repeating the process for the second sub-diagonal and $\binom{m}{2}F^2$, etc., we conclude that F must vanish identically. We note that the argument is unchanged if we replace I by λI where $|\lambda| = 1$. Hence by induction, we conclude that A itself must be diagonal. We now apply Lemma 7.1 and the result follows.

8. Representations of Finite Groups

Let $\mathfrak{g}$ be a finite group with g elements, and let $f(\xi)$ be a complex-valued function defined for ξ in $\mathfrak{g}$. The function $f(\xi)$ thus takes on exactly g values, and for each f the quantity

$$(f(\xi_1), f(\xi_2), \cdots, f(\xi_g))$$

may be considered as a vector in a g-dimensional vector space over the complex number field.

Let $U(\xi) = \|u_{ik}(\xi)\|$ be an $n \times n$ matrix which is a representation of $\mathfrak{g}$. If $L(U)$ is a linear function on matrices, then $L(U)$ has the form

$$L(U) = \sum_{i,j=1}^{n} a^{ij} u_{ij}.$$

Substituting $U = U(\xi)$ we obtain a complex-valued function defined on $\mathfrak{g}$, an element in the space of all functions defined on $\mathfrak{g}$. The totality $\{L\}$ of functions defined in this way itself forms a linear space which is a subspace of the space of all functions defined on $\mathfrak{g}$. Hence each representation $\xi \longrightarrow U(\xi)$ generates a subspace $\{L\}$, the space derived from all linear functions on matrices.

THEOREM 8.1. *Let $\xi \longrightarrow U(\xi)$ be a representation of $\mathfrak{g}$, and let C be an arbitrary non-singular matrix. Denote by $\{L\}$ the subspace of all functions generated by U. Then the representation $V(\xi) = C\,U(\xi)\,C^{-1}$ generates the same subspace as U does.*

Proof. The elements v_{ij} of the matrix $V(\xi)$ are linear homogeneous functions of the $\{u_{ik}\}$ and vice versa. Hence each v_{ij} is an element of the space $\{L\}$ belonging to U, and vice versa.

We now state the principal results; the proofs will follow.

THEOREM 8.2. *Let $U_1, U_2, \cdots, U_k$ be irreducible representations of a finite group $\mathfrak{g}$, no two of which are similar.*

Let $\mathbf{L}_1, \mathbf{L}_2, \cdots, \mathbf{L}_k$ be the subspaces of functions on $\mathfrak{g}$ generated by $U_1, U_2, \cdots, U_k$, respectively. Then $\mathbf{L}_1, \mathbf{L}_2, \cdots, L_k$ are linearly independent.

THEOREM 8.3. *Let $U(\boldsymbol{\xi})$ be an irreducible representation of $\mathfrak{g}$ of degree n. Let $\mathbf{L}$ be the subspace of functions on $\mathfrak{g}$ generated by U. Then $\mathbf{L}$ has dimension n^2.*

The proofs of Theorems 8.2 and 8.3 will depend on several lemmas that will first be established. However, before proceeding to these, several immediate consequences of Theorems 8.2 and 8.3 will be noted.

Suppose the quantities $U_1, U_2, \cdots, U_k$ of Theorem 8.2 are of dimensions $n_1, n_2, \cdots, n_k$, respectively. Then by Theorem 8.3, the spaces $\mathbf{L}_1, \mathbf{L}_2, \cdots, \mathbf{L}_k$ have dimensions $n_1^2, n_2^2, \cdots, n_k^2$. If g is the number of elements in $\mathfrak{g}$, then the space of all functions defined on $\mathfrak{g}$ is of dimension g, so we have this corollary.

COROLLARY 8.4. $n_1^2 + n_2^2 + \cdots + n_k^2 \le g.$

In particular, $k \le g$, so we can state the following result.

COROLLARY 8.5. *There are at most g irreducible representations of $\mathfrak{g}$, no two of which are similar.*

Let $U_1, U_2, \cdots, U_l$ be a system of irreducible representations that are mutually non-similar. If each irreducible representation of $\mathfrak{g}$ is similar to one of the U_i, we say that the system $U_1, U_2, \cdots, U_l$ is a **representative system** of representations. In addition to the result $l \leq g$, of Corollary 8.3, we have the following theorem.

THEOREM 8.6. *If $U_1, U_2, \cdots, U_l$ is a representative system of degrees $n_1, n_2, \cdots, n_l$, respectively, then*

$$n_1{}^2 + n_2{}^2 + \cdots + n_l{}^2 = g.$$

LEMMA 8.7. (I. Schur) *Let Σ be a collection of transformations of an m-dimensional space $\mathbf{L}$ into itself. Let $\mathbf{T}$ be a collection of transformations of an n-dimensional space $\mathbf{M}$ into itself. Let $A \neq 0$ be a linear transformation of $\mathbf{M}$ into $\mathbf{L}$. Suppose that for each transformation U in Σ there exists an element V in $\mathbf{T}$ such that the mappings UA and AV are identical, and similarly for each V in $\mathbf{T}$ there is a U in Σ such that $AV = UA$. Suppose furthermore that Σ and $\mathbf{T}$ are irreducible. Then $m = n$ and A is non-singular.*

Proof. Let y be an element of $\mathbf{M}$. Then Ay is in $\mathbf{L}$. We denote the totality of vectors Ay, by $A(\mathbf{M})$. Then $A(\mathbf{M})$ is a linear subspace of $\mathbf{L}$, which we shall show is invariant under Σ. For let U be an element of Σ and consider UAy. By hypothesis, there exists a V in $\mathbf{T}$ such that $UAy = AVy$. But Vy is an element of $\mathbf{M}$, hence $A(Vy)$ is in $A(\mathbf{M})$. Thus U maps $A(\mathbf{M})$ into $A(\mathbf{M})$. Since Σ is irreducible and A is not zero, $A(\mathbf{M})$ is the entire space $\mathbf{L}$. This implies that $m \leq n$ since a linear mapping cannot increase dimension. By considering the transposes of the matrices A, U, V, we have $A'U' = V'A'$. Interchanging the roles of the spaces $\mathbf{M}$ and $\mathbf{L}$, we find $n \leq m$ by the same argument. Hence $m = n$. Thus the mapping A is onto; but this can occur only if A is non-singular.

COROLLARY 8.8. *Under the hypotheses of Lemma 8.7, the collections Σ and $\mathbf{T}$ are similar.*

For A is non-singular; hence the relation $UA = AV$ may be written $V = A^{-1}UA$.

LEMMA 8.9. *Let U and V be irreducible non-similar representations in the spaces $\mathbf{L}$ and $\mathbf{M}$, respectively. Let C be an arbitrary linear transformation of $\mathbf{M}$ into $\mathbf{L}$. Then the transformation A defined by*

$$A = \frac{1}{g} \sum_{\xi \in \mathfrak{g}} U(\xi) C V^{-1}(\xi)$$

is the zero transformation.

Proof. We first note that for ξ in $\mathfrak{g}$, the relation $V^{-1}(\xi) = V(\xi^{-1})$ holds.

We form the expression

$$U(\alpha)\,A\,V^{-1}(\alpha) = \frac{1}{g}\,U(\alpha)\left[\sum_{\xi} U(\xi)\,C\,V^{-1}(\xi)\right]V^{-1}(\alpha)$$

$$= \frac{1}{g}\sum_{\xi} U(\alpha\xi)\,C\,V^{-1}(\xi)\,V^{-1}(\alpha) = \frac{1}{g}\sum_{\xi} U(\alpha\xi)\,C\,V(\xi^{-1})\,V(\alpha^{-1})$$

$$= \frac{1}{g}\sum_{\xi} U(\alpha\xi)\,C\,V(\xi^{-1}\alpha^{-1}) = \frac{1}{g}\sum_{\xi} U(\alpha\xi)\,C\,V[(\alpha\xi)^{-1}]$$

$$= \frac{1}{g}\sum_{\xi} U(\alpha\xi)\,C\,V^{-1}(\alpha\xi) = A.$$

Hence $U(\alpha)\,A\,V^{-1}(\alpha) = A$ for all α in $\mathfrak{g}$. If A were not zero, then the hypotheses of Lemma 8.7 would hold, interpreting the collection of transformations $U(\alpha)$ as Σ and that of the $V(\alpha)$ as $\mathbf{T}$. (These sets are irreducible since by hypothesis the representations U, V are irreducible.) Hence by Corollary 8.8, U and V are similar, which is impossible also by hypothesis. Therefore $A = 0$.

LECTURE IV

9. Representations of Finite Groups (continued)

We observed (Theorem 5.5) that any representation of a finite group is similar to a unitary representation. Thus we may assume $U(\xi)$ and $V(\xi)$ are (irreducible, non-similar) unitary representations of the finite group $\mathfrak{g}$ on spaces **L** and **M**, respectively. The relation

$$\sum_{\xi} U(\xi)CV^{-1}(\xi) = 0$$

may be rewritten

$$\sum_{\xi} U(\xi)CV^*(\xi) = 0.$$

This is the case because each $V(\xi)$ is unitary, i.e., $V^{-1}(\xi) = V^*(\xi)$, where * denotes the transpose conjugate operation on a matrix. To write this in terms of components, we set $C = \|c_{\rho\sigma}\|$ and find

$$\sum_{\xi} \sum_{\rho,\sigma} u_{i\rho}(\xi)c_{\rho\sigma}\overline{v_{j\sigma}(\xi)} = 0.$$

Lemma 8.9 stated that this equation holds for all matrices C. In particular, if we take for C the matrix with 1 in the (k, l) position and 0 elsewhere, we obtain

$$\sum_{\xi} u_{ik}(\xi)\overline{v_{jl}(\xi)} = 0$$

for $1 \leq i, k \leq m = \dim(\mathbf{L})$, $1 \leq j, l \leq n = \dim(\mathbf{M})$. We interpret this formula by introducing a Hermitian metric on the linear space of all complex-valued functions on the group space:

$$f_1 \cdot f_2 = \frac{1}{g} \sum_\xi f_1(\xi)\overline{f_2(\xi)}.$$

Our result now has the simple form:

$$u_{ik} \cdot v_{jl} = 0.$$

We now denote by **F** the space of functions on the group space determined by the representation $U(\xi)$. We recall that this is the subspace (of the space of all functions on $\mathfrak{g}$) that consists of all functions of the form $\sum a^{ik} u_{ik}(\xi)$. That is, **F** is spanned by the functions u_{ik}. Similarly the space **G** determined by $V(\xi)$ is spanned by the functions v_{jl}, However, we proved that each u_{ik} is orthogonal to each v_{jl}. Consequently we have the following result.

LEMMA 9.1. *The spaces* **F** *and* **G** *are orthogonal. In other words, two irreducible non-similar representations determine orthogonal subspaces in the space of all functions on the group.*

We shall use the notation

$$\tau(C) = \sum c_{jj}$$

for the trace of a square matrix $C = \|c_{ij}\|$. We recall that τ is linear, $\tau(AB) = \tau(BA)$, $\tau(PAP^{-1}) = \tau(A)$.

LEMMA 9.2. *Let $U(\xi)$ be an irreducible representation of degree m of the group $\mathfrak{g}$ of g elements. Let C be any $g \times g$ matrix. Then*

$$\frac{1}{g} \sum_\xi U(\xi) C U(\xi)^{-1} = \frac{1}{m} \tau(C) I.$$

Proof. We set

$$A = \frac{1}{g} \sum_\xi U(\xi) C U(\xi)^{-1}.$$

In the proof of Lemma 8.9, we showed that for each α, $U(\alpha) A = A U(\alpha)$. Hence for each constant λ we have

$$U(\alpha) (A - \lambda I) = (A - \lambda I) U(\alpha).$$

We chose λ to be a characteristic root of A, so that $\det (A - \lambda I) = 0$. Then $A - \lambda I$ is a singular matrix which, by the Schur Lemma 8.7, must vanish; that is, $A = \lambda I$. Of course λ depends on C; to indicate this dependence we write $\lambda = \lambda(C)$, and we have proved

$$\frac{1}{g} \sum_{\xi} U(\xi) C U(\xi)^{-1} = \lambda(C) I.$$

We take traces:

$$\frac{1}{g} \sum_{\xi} \tau[U(\xi) C U(\xi)^{-1}] = \tau[\lambda(C) I],$$

$$\frac{1}{g} \sum_{\xi} \tau(C) = m\lambda(C), \qquad \tau(C) = m\lambda(C).$$

Hence, $\lambda(C) = (1/m)\tau(C)$.

Again we assume $U(\xi)$ is a unitary representation. Then our result becomes

$$\frac{1}{g} \sum_{\xi} \sum_{\rho,\sigma} u_{i\rho}(\xi) c_{\rho\sigma} \overline{u_{j\sigma}(\xi)} = \frac{1}{m} \left(\sum_{r} c_{rr} \right) \delta_{ij}.$$

Since this is true for all choices of the matrix C, we may equate coefficients of the $c_{\rho\sigma}$ to obtain

$$\frac{1}{g} \sum_{\xi} u_{i\rho}(\xi) \overline{u_{j\sigma}(\xi)} = \frac{1}{m} \delta_{ij} \delta_{\rho\sigma}.$$

From this we read off the following result.

LEMMA 9.3. *The functions $u_{i\rho}$ and $u_{j\sigma}$ are orthogonal unless $i = j$ and $\rho = \sigma$. In this case, $u_{i\rho} \cdot u_{i\rho} = 1/m$. Consequently, the functions $u_{i\rho}$ are linearly independent and the space $\mathbf{F}$ they span has dimension m^2.*

Suppose now that $U_1, \cdots, U_l$ are irreducible, non-similar representations of degrees $m_1, \cdots, m_l$, respectively. Each U_i determines a subspace $\mathbf{F}_i$ of the g-dimensional space (with Hermitian metric) of all functions on the group. We have seen that $\dim (\mathbf{F}_i) = m_i^2$ and that the $\mathbf{F}_i$ are mutually orthogonal, hence linearly independent. The space they span must have dimension $\sum m_i^2$; consequently

$$\sum_{1}^{l} m_i^2 \leq g,$$

and in particular $l \leq g$.

We shall call $U_1, \cdots, U_l$ a **complete** system of representations if each irreducible representation is similar to one of them. We may assume that we have such a system.

We note the following fact. Let $U(\xi)$ be any representation of $\mathfrak{g}$, irreducible or not. We have proved that $U(\xi)$ is completely reducible, hence we may decompose U into irreducible parts

$$U = \begin{pmatrix} U' & & & \\ & U'' & & \\ & & \cdot & \\ & & & \cdot \\ & & & & \cdot \end{pmatrix}.$$

If **F** is the space of functions associated with U, it is clear that the totality of functions $u'_{ij}, u''_{kl}, \cdots$ spans **F**. But the functions u'_{ij}, coming from an irreducible representation, span one of the spaces $\mathbf{F}_i$. The same holds for the functions u''_{kl}, etc. Consequently the space **F** is the sum of certain of the spaces $\mathbf{F}_1, \cdots, \mathbf{F}_l$ associated with a complete system of irreducible representations.

We apply this in particular to the **Cayley representation** by right translations. Each α in $\mathfrak{g}$ defines a linear transformation $R(\alpha)$ on the g-dimensional space of functions on $\mathfrak{g}$:

$$R_\alpha : f(\xi) \longrightarrow f(\xi\alpha).$$

We easily check that $R(\alpha\beta) = R(\alpha) R(\beta)$.

LEMMA 9.4. *The space* **F** *associated with the Cayley representation is the total space of functions on* $\mathfrak{g}$.

To prove this, we write

$$\mathfrak{g} = \{\alpha_1 = \epsilon, \alpha_2, \cdots, \alpha_g\},$$

which leads to a basis $f_1, \cdots, f_g$ of the space of functions, defined by

$$f_i(\alpha_j) = \delta_{ij}.$$

Thus, in coordinate notation,

$$f_1 = \begin{bmatrix} 1 \\ 0 \\ 0 \\ \cdot \\ \cdot \\ \cdot \end{bmatrix}, \quad f_2 = \begin{bmatrix} 0 \\ 1 \\ 0 \\ \cdot \\ \cdot \\ \cdot \end{bmatrix}, \cdots.$$

We shall compute with respect to this basis the first column of the matrix $U(\alpha_i)$ associated with $R(\alpha_i)$. First of all we note the relation

$$U(\alpha_i) f_j = f_k \quad \text{where} \quad \alpha_k = \alpha_j \alpha_i^{-1}.$$

For

$$[U(\alpha_i)f_j](\alpha) = f_j(\alpha\alpha_i) = \begin{cases} 1 & \text{if } \alpha_j = \alpha\alpha_i \\ 0 & \text{if } \alpha_j \neq \alpha\alpha_i. \end{cases}$$

But the relation $\alpha_j = \alpha\alpha_i$ is the same as $\alpha_j\alpha_i^{-1} = \alpha$. Consequently, if we write

$$U(\alpha_i)f_j = \sum_\rho u_{j\rho}(\alpha_i)f_\rho,$$

then $u_{j1}(\alpha_i) = 0$ for $i \neq j$ and $u_{i1}(\alpha_i) = 1$. It follows that the first column of $U(\alpha_i)$ is precisely f_i. This implies in turn that the functions u_{j1} $(j = 1, \cdots, g)$ alone span the total space of functions on $\mathfrak{g}$.

COROLLARY 9.5. *If $U_1, \cdots, U_l$ is a complete system of irreducible representations of degrees $m_1, \cdots, m_l$ respectively, if each U_i determines the space $\mathbf{F}_i$, and if $\mathbf{F}$ is the total space of functions on $\mathfrak{g}$, then*

$$\mathbf{F} = \mathbf{F}_1 + \cdots + \mathbf{F}_l$$

and

$$g = m_1{}^2 + \cdots + m_l{}^2.$$

For the Cayley representation determines the total space $\mathbf{F}$. But we observed that the space determined by an arbitrary representation is a sum of several of the $\mathbf{F}_i$. The result follows from this fact combined with the linear independence of the $\mathbf{F}_i$.

10. Characters

DEFINITION 10.1. Let $U(\alpha)$ be a representation of $\mathfrak{g}$. The **character** of the representation is the function

$$\chi(\alpha) = \tau[U(\alpha)].$$

obtained by taking the trace. A **primitive character** is the character of an irreducible representation.

We note immediately that *similar representations have the same character* since similar matrices have equal traces. We next observe that *the character of a representation determines the characteristic polynomial of each $U(\alpha)$.* For the identities of Newton express the elementary symmetric functions of the characteristic roots of $U(\alpha)$ in terms of the traces of the successive powers of this matrix. But

$$\tau[U^2(\alpha)] = \tau[U(\alpha^2)] = \chi(\alpha^2),$$

$$\tau[U^3(\alpha)] = \tau[U(\alpha^3)] = \chi(\alpha^3), \quad \text{etc.}$$

If $U_1(\alpha)$, $U_2(\alpha)$ are non-similar irreducible representations, then their corresponding primitive characters χ_1, χ_2 are orthogonal, since they belong, respectively, to the orthogonal subspaces $\mathbf{F}_1$, $\mathbf{F}_2$ associated with U_1, U_2.

Elements α, β of $\mathfrak{g}$ are called **conjugate** if there is a γ in $\mathfrak{g}$ such that $\alpha = \gamma\beta\gamma^{-1}$. Thus we introduce an equivalence relation on $\mathfrak{g}$, which decomposes $\mathfrak{g}$ into **conjugate classes.** A function f on $\mathfrak{g}$ which is constant on each conjugate class is called a **class function.** We see that *each character of a representation is a class function.* For $\chi(\alpha) = \chi(\gamma\beta\gamma^{-1}) = \tau[U(\gamma\beta\gamma^{-1})] = \tau[U(\gamma)U(\beta)U^{-1}(\gamma)] = \tau[U(\beta)] = \chi(\beta)$.

LEMMA 10.2. *The primitive characters span the space of class functions.*

Proof. Let $U_1, \cdots, U_l$ be a complete system with corresponding primitive characters $\chi_1, \cdots, \chi_l$. If f is any function on $\mathfrak{g}$, then by our previous result

$$f(\xi) = \sum a_{ij}^1 u_{ij}^1(\xi) + \sum a_{ij}^2 u_{ij}^2(\xi) + \cdots$$
$$= \tau[A_1 U_1(\xi)] + \tau[A_2 U_2(\xi)] + \cdots,$$

where $A_1 = \|a_{ji}^1\|$, $A_2 = \|a_{ji}^2\|$, $\cdots$. If f is a class function, then $f(\xi) = f(\alpha\xi\alpha^{-1})$ for each α, hence

$$f(\xi) = \tau[A_1 U_1(\alpha\xi\alpha^{-1})] + \cdots$$
$$= \tau[A_1 U_1(\alpha)U_1(\xi)U_1^{-1}(\alpha)] + \cdots$$
$$= \tau[U_1^{-1}(\alpha)A_1 U_1(\alpha)U_1(\xi)] + \cdots.$$

We sum this relation on α to obtain

$$gf(\xi) = \tau\left\{\left[\sum_\alpha U_1^{-1}(\alpha)A_1 U_1(\alpha)\right]U_1(\xi)\right\} + \cdots$$
$$= \tau\{[c_1 I_1]U_1(\xi)\} + \cdots$$
$$= c_1\chi_1(\xi) + \cdots.$$

We have used Lemma 9.2 to evaluate the sum on α. It follows that $f = c_1\chi_1 + \cdots + c_l\chi_l$ as asserted.

THEOREM 10.3. *The number l of non-similar irreducible representations is equal to the number of conjugate classes in $\mathfrak{g}$.*

Proof. The primitive characters are non-zero (since, for example, $\chi_i(\epsilon) = m_i =$ degree of $U_i \neq 0$), and they are mutually orthogonal, hence linearly independent. Consequently, the space they span has dimension l. But this space is the space of class functions; its dimension is evidently equal to the number of classes.

In particular, we have $l = g$ only in case $\mathfrak{g}$ is an abelian group.

LECTURE V

11. Representations of Finite Groups (concluded)

Let $U(\alpha)$ be an irreducible (unitary) representation of degree m of the finite group $\mathfrak{g}$ of g elements. In the last lecture we deduced the basic relations (Lemma 9.3),

$$\frac{1}{g} \sum_{\xi} u_{ik}(\xi) \overline{u_{jl}(\xi)} = \frac{1}{m} \delta_{ij} \delta_{kl},$$

from which we obtain as a special case

$$\frac{1}{g} \sum_{\xi} u_{ii}(\xi) \overline{u_{ii}(\xi)} = \frac{1}{m}.$$

The character χ of the representation is given by

$$\chi(\alpha) = \sum_{i} u_{ii}(\alpha),$$

and hence

$$\chi \cdot \chi = \frac{1}{g} \sum_{i,j} \sum_{\xi} u_{ii}(\xi) \overline{u_{jj}(\xi)} = \frac{1}{g} \sum_{i} \sum_{\xi} u_{ii}(\xi) \overline{u_{ii}(\xi)} = \frac{1}{g} \sum_{i} \frac{g}{m} = 1.$$

We have proved the following result.

LEMMA 11.1. *If $\chi(\alpha)$ is the character of an irreducible representation $U(\alpha)$, then $\chi \cdot \chi = 1$.*

Now suppose $U(\alpha)$ is any representation of the group $\mathfrak{g}$, irreducible or not. We decompose $U(\alpha)$ into irreducible parts:

$$U(\alpha) = \begin{bmatrix} U'(\alpha) & & & \\ & U''(\alpha) & & \\ & & \cdot & \\ & & & \cdot \\ & & & & \cdot \end{bmatrix}.$$

For the corresponding characters, we have

$$\chi_U(\alpha) = \chi_{U'}(\alpha) + \chi_{U''}(\alpha) + \cdots.$$

Now let $U_1, \cdots, U_l$ be a representative system for the irreducible representations of $\mathfrak{g}$. Each U_i occurs a certain number of times in the decomposition of U. Let us write n_i for the number of times U_i occurs, and denote by χ_i the character for U_i. Thus

$$\chi_U(\alpha) = n_1\chi_1(\alpha) + \cdots + n_l\chi_l(\alpha).$$

Since the distinct χ_i are orthogonal, we immediately obtain

$$\chi_U \cdot \chi_U = n_1{}^2 + n_2{}^2 + \cdots + n_l{}^2.$$

From this we deduce the following theorem.

THEOREM 11.2. *The representation $U(\alpha)$ with character $\chi(\alpha)$ is irreducible if and only if $\chi \cdot \chi = 1$.*

Proof. We have already seen that $\chi \cdot \chi = 1$ if $U(\alpha)$ is irreducible. If on the other hand $U(\alpha)$ is reducible, then either some n_i exceeds 1, in which case $\chi \cdot \chi = \sum n_i{}^2 > 1$, or at least two of the n_i are non-zero, in which case $\chi \cdot \chi = \sum n_i{}^2 \geq 2$.

Suppose we have another decomposition of $U(\alpha)$:

$$CU(\alpha)C^{-1} = \begin{bmatrix} V'(\alpha) & & & \\ & V''(\alpha) & & \\ & & \cdot & \\ & & & \cdot \\ & & & & \cdot \end{bmatrix}.$$

If each $U_i(\alpha)$ appears n_i' times, we have

$$\chi_U(\alpha) = n_1'\chi_1(\alpha) + \cdots + n_l'\chi_l(\alpha)$$

for the character of $U(\alpha)$. From this expression and the previous expression for χ_U, we have

$$(n_1 - n_1')\chi_1(\alpha) + \cdots + (n_l - n_l')\chi_l(\alpha) = 0.$$

But the χ_α are linearly independent, and hence the coefficients vanish. Therefore $n_1 = n_1', n_2 = n_2', \cdots$. We have established the following result.

Theorem 11.3. *In two decompositions of a representation into irreducible parts, each irreducible representation (of a complete system) occurs exactly the same number of times. Hence there is essentially but one decomposition.*

By looking at the last argument from a slightly different point of view, we also get the following result.

Theorem 11.4. *Two representations are equivalent if and only if they have the same character.*

We complete our study of finite groups with a further analysis of the regular, or Cayley, representation. We recall that this is the representation of $\mathfrak{g}$ on the g-dimensional space of (complex-valued) functions on the group space given by

$$f(\xi) \longrightarrow f(\xi\alpha).$$

It is a representation of degree g. We proved (Corollary 9.5) that if $U_1, \cdots, U_l$ is a complete system of irreducible representations, and if each U_i is of degree n_i, then

$$n_1^2 + n_2^2 + \cdots + n_l^2 = g.$$

Let us denote by m_i the number of times U_i occurs in the Cayley representation. Since the typical matrix of the Cayley representation is a $g \times g$ matrix, and since it can be represented as a matrix of diagonal blocks, with U_1 appearing m_1 times, etc., it follows by counting rows that

$$n_1 m_1 + n_2 m_2 + \cdots + n_l m_l = g.$$

Now let χ denote the character of the Cayley representation. We proved (indeed, for any representation) that $\chi \cdot \chi = m_1^2 + \cdots + m_l^2$.

Lemma 11.5. *We have $\chi \cdot \chi = g$.*

Proof. In the Cayley representation, the mapping $\xi \longrightarrow \xi\alpha$ is a permutation of the elements of $\mathfrak{g}$. If $\alpha \neq \epsilon$, then each element ξ is moved under this permutation, that is, all entries in the principal diagonal of the corresponding matrix vanish. Thus $\chi(\alpha) = 0$ if $\alpha \neq \epsilon$. But $\chi(\epsilon) = \tau(I) = g$. Finally,

$$\chi \cdot \chi = \frac{1}{g} \sum_\xi \chi(\xi) \cdot \overline{\chi(\xi)} = \frac{1}{g} (g^2) = g.$$

We have obtained three relations:

$$\sum n_i{}^2 = g, \quad \sum m_i n_i = g, \quad \sum m_i{}^2 = g.$$

From these we easily deduce that $n_i = m_i$, either by noting that the vectors $\mathbf{m} = (m_1, \cdots, m_l)$ and $\mathbf{n} = (n_1, \cdots, n_l)$ satisfy $\mathbf{m} \cdot \mathbf{m} = \mathbf{m} \cdot \mathbf{n} = \mathbf{n} \cdot \mathbf{n}$ and hence are equal by elementary geometry; or by adding the first and third sums above and subtracting twice the second to obtain $\sum (m_i - n_i)^2 = 0$.

THEOREM 11.6. *In the Cayley representation, each irreducible $U_i(\alpha)$ occurs precisely n_i times, where n_i is the degree of U_i.*

12. Introduction to Differentiable Manifolds

We begin by reviewing the notion of a topological space.

DEFINITION 12.1. A **Hausdorff space** is set a $\mathscr{S}$ in which to each point p are associated certain sets $\mathscr{N}_1(p), \mathscr{N}_2(p), \cdots$, called **neighborhoods** of p, which satisfy the following axioms:

1. $p \in \mathscr{N}(p)$

2. If $\mathscr{N}_1$ and $\mathscr{N}_2$ are neighborhoods of p, then there exists a third neighborhood $\mathscr{N}$ of p such that $\mathscr{N} \subset \mathscr{N}_1 \cap \mathscr{N}_2$.

3. If $q \in \mathscr{N}(p)$, then there is a neighborhood $\mathscr{N}(q)$ of q such that $\mathscr{N}(q) \subset \mathscr{N}(p)$.

4. If $q_1 \neq q_2$, then there are neighborhoods $\mathscr{N}(q_1)$ and $\mathscr{N}(q_2)$ such that $\mathscr{N}(q_1) \cap \mathscr{N}(q_2) = \varnothing$.

We remark that the notions of *open set, closed set,* and *continuous function* are defined on a Hausdorff space. Next we set down the definition of a topological group.

DEFINITION 12.2. A **topological group** is a group $\mathfrak{g}$ on which there is a Hausdorff topology such that the operations

$$(\sigma, \tau) \longrightarrow \sigma\tau \quad \text{and} \quad \tau \longrightarrow \tau^{-1}$$

are continuous. We also call $\mathfrak{g}$ a **continuous group**.

We note that the two conditions on continuity of the group operations may be replaced by a single condition, namely that $(\sigma, \tau) \longrightarrow \sigma\tau^{-1}$ is continuous. For if this is the case, then taking $\sigma = \epsilon$ yields the continuity of $\tau \longrightarrow \tau^{-1}$ and hence of $(\sigma, \tau) \longrightarrow \sigma\tau = \sigma(\tau^{-1})^{-1}$.

We shall restrict our study to those continuous groups for which the group operations are not only continuous but are differentiable functions. To formulate precisely what is meant, we need the concept of a differentiable manifold.

A **differentiable manifold** of **order** k and **dimension** n consists of a Hausdorff space $\mathscr{S}$ and a set $\sum$ of real functions, each defined on some open set $\mathcal{O}$ of $\mathscr{S}$. We require the following postulates:

1. Let $f \in \sum$ and f be defined on $\mathcal{O}$. Suppose $\mathcal{O}_1 \subset \mathcal{O}$. Then f considered only on $\mathcal{O}_1$ is also in $\sum$.

2. Let $f_1, \cdots, f_r \in \sum$, all defined on $\mathcal{O}$. Consider the mapping $x_1 = f_1(p), \cdots, x_r = f_r(p)$ of $\mathcal{O}$ into r-dimensional Euclidean space $\mathscr{E}_r$. Let Γ be the set thus obtained in $\mathscr{E}_r$. Suppose that Ω is an open domain in $\mathscr{E}_r$, that $\Gamma \subset \Omega$, and that $F(x_1, \cdots, x_r)$ is a k-times continuously differentiable function on Ω. Then, considered as a function on $\mathcal{O}$, $F(f_1(p), \cdots, f_r(p)) \in \sum$.

3. There is a system of open sets $\mathcal{O}$ which cover $\mathscr{S}$ such that each $\mathcal{O}$ of the system can be mapped homeomorphically onto a sphere Ω of $\mathscr{E}_n$ by a mapping of the form $x_1 = f_1(p), \cdots, x_n = f_n(p)$ with $f_i \in \sum$; and such that if $\mathcal{O}_1 \subset \mathcal{O}$, then any function of $\sum$ defined on $\mathcal{O}_1$ can be represented as a k-times continuously differentiable function (i.e., of class $C^{(k)}$) of $x_1, \cdots, x_n$ on the image Ω_1 of $\mathcal{O}_1$, and conversely.

We also call $\mathscr{S}$ a **smooth manifold of order** k or a k**-times differentiable manifold.**

NOTE. There exist Hausdorff spaces which cannot be made into manifolds. For example, the cross in the Euclidean plane (Figure 1). For if a neighborhood of the origin were homeomorphic to an interval, then removal of a single point would disconnect the cross into only two components. However, removal of the origin makes four components.

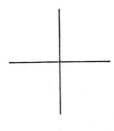

FIGURE I

DEFINITION 12.3. Let $\mathcal{O}^*$ be an open set on $\mathscr{S}$ and let $l \leq k$. A function $f(p)$ defined on $\mathcal{O}^*$ is said to be of **class** $C^{(l)}$ (l**-times continuously differentiable** provided that given any $p_0 \in \mathscr{O}^*$, there is one of the covering neighborhoods $\mathscr{O}$ of Postulate 3 above containing p_0 such that in a sufficiently small neighborhood of p_0, $f(p) = F(x_1, \cdots, x_n)$, where $F(x_1, \cdots, x_n)$ is of class $C^{(k)}$ on Ω. (Here $x_i = f_i(p)$ defines the homeomorphism of $\mathscr{O}$ onto a euclidean sphere Ω.)

We note that there is a consistency in this definition: it does not matter which of the covering neighborhoods is chosen. In fact let $\mathscr{O}_1$ be another of the neighborhoods of the definition such that $p \in \mathscr{O}_1$, with corresponding coordinate functions $y_1, \cdots, y_n$ mapping $\mathscr{O}_1$ homeomorphically onto a sphere Ω_1. Then on the common part $\mathscr{O}_2 = \mathscr{O} \cap \mathscr{O}_1$, also an open set, each x_i is a $C^{(k)}$ function of the y_j. Thus the function $F(x_1, \cdots, x_n)$ is a $C^{(l)}$ function of the variables $y_1, \cdots, y_n$.

We also note in this situation that the relation

$$\left\| \frac{\partial y_i}{\partial x_j} \right\| \cdot \left\| \frac{\partial x_i}{\partial y_j} \right\| = I$$

holds between the Jacobian matrices, and that the consequent relation

$$J\!\left(\begin{matrix} y \\ x \end{matrix}\right) \cdot J\!\left(\begin{matrix} x \\ y \end{matrix}\right) = 1$$

holds between the Jacobians, from which we conclude that

$$J\!\left(\begin{matrix} y \\ x \end{matrix}\right) \neq 0.$$

DEFINITION 12.4. A **coordinate system** in an open set $\mathscr{O}$ of $\mathscr{S}$ is a set of n functions $y_1 = f_1(p), \cdots, y_n = f_n(p)$ defined on $\mathscr{O}$, each of class $C^{(k)}$ on $\mathscr{O}$, with the following two properties:

1. If $p \in \mathscr{O}$, and $x_1, \cdots, x_n$ are the functions on a covering neighborhood $\mathscr{O}_1$ in the definition of smooth manifold, so that $f_i(p) = F_i(x_1, \cdots, x_n)$ on $\mathscr{O}_1 \cap \mathscr{O}$, then

$$\left| \frac{\partial F_i}{\partial x_j} \right| \neq 0 \quad \text{on} \quad \mathscr{O}_1 \cap \mathscr{O}.$$

2. The mapping $p \longrightarrow (y_1, \cdots, y_n)$ is a homeomorphism on $\mathscr{O}$ onto an open set Ω of E_n.

This definition is independent of the coordinate functions x_i. For if x_i' are new coordinates, then

$$\left| \frac{\partial F_i}{\partial x_j'} \right| = \left| \frac{\partial F_i}{\partial x_j} \right| \cdot \left| \frac{\partial x_j}{\partial x_i'} \right| \neq 0,$$

by the well-known rule for multiplying Jacobians.

13. Tensor Calculus on a Manifold

We shall work on an n-dimensional manifold $\mathscr{S}$ of order $k > 0$. Suppose that $p_0 \in \mathscr{S}$ and that $x^1, \cdots, x^n; y^1, \cdots, y^n$ are coordinate systems in a neighborhood $\mathscr{O}$ of p_0. We suppose the coordinates of p_0 are $x_0{}^i$ and $y_0{}^i$ in the respective systems. The differences $\Delta y^i = y^i - y_0{}^i$, up to first-order terms, are given by

$$\Delta y^i = \sum_\rho \left(\frac{\partial y^i}{\partial x^\rho}\right)_{x_0} \Delta x^\rho.$$

Thus each transition from one coordinate system at p_0 to another has associated with it a certain non-singular linear transformation. What we have essentially is the transformation law for differentials:

$$dy^i = \sum_\rho \left(\frac{\partial y^i}{\partial x^\rho}\right)_{p_0} dx^\rho.$$

If z^i is yet another coordinate system, then

$$dz^i = \sum_\rho \left(\frac{\partial z^i}{\partial y^\rho}\right)_{p_0} dy^\rho$$

If we compose the transformations, we obtain

$$dz^i = \sum_\rho \left(\frac{\partial z^i}{\partial x^\rho}\right)_{p_0} dx^\rho,$$

which yields

$$\left\| \frac{\partial z^i}{\partial x^\rho} \right\| = \left\| \frac{\partial z^i}{\partial y^\sigma} \right\| \cdot \left\| \frac{\partial y^\sigma}{\partial x^\rho} \right\|.$$

This shows that we have a homomorphism and allows us to define a **tangent vector** of $\mathscr{S}$ at p_0. This is given, up to a choice of coordinate system x^i around p_0, by an ordered system of n numbers $a^1, \cdots, a^n$. If we pass to the coordinate system y^i, the a^j must be replaced by $b^1, \cdots, b^n$ obtained from the transformation leading from dx^i to dy^i according to the rule

$$b^i = \sum_\rho \left(\frac{dy^i}{dx^\rho}\right)_{p_0} a^\rho.$$

The definition is unambiguous, because if z^i is a third coordinate system and

$$c^i = \sum_\rho \left(\frac{\partial z^i}{\partial y^\rho}\right)_{p_0} b^\rho,$$

then by composing transformations we obtain

$$c^i = \sum_\rho \left(\frac{\partial z^i}{\partial x^\rho}\right)_{p_0} a^\rho.$$

We shall also refer to a tangent vector as a **contravariant vector**.

A contravariant vector forms the simplest type of what we shall call a quantity on $\mathscr{S}$. In general a **quantity** is defined by numbers $u_1, \cdots, u_p$ in each coordinate system x^i, such that the transition from the system x^i to a new coordinate system y^i leads to the numbers $v_1, \cdots, v_p$ according to a definite rule of the form

$$v_i = F_i\left(\frac{\partial(y)}{\partial(x)}, \frac{\partial^2(y)}{\partial(x)^2}, \cdots, \frac{\partial^k(y)}{\partial(x)^k}, \mathbf{u}\right)_{p_0}.$$

We insist on a consistency relation: if $w_1, \cdots, w_p$ are the components of the quantity in a third coordinate system z_i, then

$$w_i = F_i\left(\frac{\partial(z)}{\partial(y)}, \cdots, \frac{\partial^k(z)}{\partial(y)^k}, \mathbf{v}\right)_{p_0}$$

with the *same* functions F_i.

LECTURE VI

14. Quantities, Vectors, and Tensors

Let $\mathbf{x} = (x^1, x^2, \cdots, x^n) \longrightarrow \mathbf{y} = (y^1, y^2, \cdots, y^n)$ be a transformation of coordinates in the neighborhood of a point P of an n-dimensional manifold, smooth at least of order l. We consider **elements** $\boldsymbol{\alpha}$ of the form

$$\boldsymbol{\alpha} = \left(\frac{\partial y^1}{\partial x^1}, \frac{\partial y^1}{\partial x^2}, \cdots \frac{\partial y^n}{\partial x^n}, \frac{\partial^2 y^1}{(\partial x^1)^2}, \cdots, \frac{\partial^l y^n}{(\partial x^n)^l} \right)_P,$$

which consist of all the partial derivatives of the y^i with respect to the x^j up to order l, evaluated at the point P. The element $\boldsymbol{\alpha}$ characterizes the transformation at P up to the order l. If $y \longrightarrow z$ is the transformation to a third system of coordinates, we form the element

$$\boldsymbol{\beta} = \left(\frac{\partial z}{\partial y}, \frac{\partial^2 z}{(\partial y)^2}, \cdots, \frac{\partial^l z}{(\partial y)^l} \right)_P$$

in the same way. If we consider the composed transformation $\mathbf{x} \longrightarrow \mathbf{z}$ at P, an element $\boldsymbol{\gamma}$ is obtained, which evidently can be computed from $\boldsymbol{\alpha}$ and $\boldsymbol{\beta}$ by the rules for differentiation of composite functions. We write $\boldsymbol{\gamma} = \boldsymbol{\beta}\boldsymbol{\alpha}$. We may now consider the $\boldsymbol{\alpha}, \boldsymbol{\beta}, \boldsymbol{\gamma}, \cdots$ as elements, the components of which are **free constants** (i.e., completely arbitrary) restricted only by the requirement $J(\frac{y}{x}) \neq 0$. We see in virtue of the composition rule for these elements that they define a group. We call this group $\mathfrak{G}_{n,l}$.

We shall now define a **quantity** of order l on a manifold of dimension n and smoothness at least l. Our point of departure is a realization of $\mathfrak{G}_{n,l}$ by a group of transformations in a space which we assume to be Euclidean, say of dimension p. To each element $\boldsymbol{\alpha}$ of $\mathfrak{G}_{n,l}$ is associated a transformation τ_α

transforming a point $\mathbf{u} = (u_1, u_2, \cdots, u_p)$ of the Euclidean space into a point $\mathbf{v} = (v_1, v_2, \cdots, v_p)$ by the rule

$$v_i = F_i(\mathbf{u}, \boldsymbol{\alpha}), \qquad i = 1, 2, \cdots, p.$$

The transformations τ_α satisfy the fundamental homomorphism requirement

$$\tau_{\alpha\beta} = \tau_\alpha \tau_\beta.$$

With each such realization of $\mathfrak{G}_{n,l}$ is now associated a quantity according to the following definition: If a coordinate system x is introduced in the neighborhood of a point P, a **quantity** at P is described by a system of values $\mathbf{u} = (u_1, u_2, \cdots, u_p)$, which may be considered as coordinates of the quantity. If a new coordinate system y is introduced in a neighborhood of P, the coordinates of the quantity should transform according to the rule

$$v_i = F_i(\mathbf{u}, \boldsymbol{\alpha}), \qquad i = 1, 2, \cdots, p.$$

The homomorphism requirement assures us that no contradiction will occur.

Examples. As we saw above, a contravariant vector is a quantity described by the relation

$$v^i = \sum_\rho \frac{\partial y^i}{\partial x^\rho} u^\rho,$$

for a transformation $x \longrightarrow y$.

A **covariant vector** is a quantity $(u_1, u_2, \cdots, u_p)$ characterized by the condition that for each contravariant vector $u^1, u^2, \cdots, u^p$, the scalar $\sum_{i=1}^p u_i u^i$ is an invariant. This implies that the components of a covariant vector obey the transformation law

$$u_i = \sum_{\rho=1}^p \frac{\partial y^\rho}{\partial x^i} v_\rho, \quad \text{or} \quad v_i = \sum_{\rho=1}^p \frac{\partial x^\rho}{\partial y^i} u_\rho.$$

Let

$$u' = [(u')^1, \cdots, (u')^p], \qquad u'' = [(u'')^1, \cdots, (u'')^p],$$

be contravariant vectors. The quantity $u^{ik} = (u')^i (u'')^k$ is a **contravariant tensor** of the second order. The transformation law is given by

$$v^{ik} = (v')^i (v'')^k = \sum_{\rho,\sigma} \frac{\partial y^i}{\partial x^\rho} \frac{\partial y^k}{\partial x^\sigma} u^{\rho\sigma}.$$

Of course not all second order tensors are the products of vectors. In a similar way, **covariant tensors** of the second order may be generated by the products $u_i' \, u_k''$, and **mixed tensors** of the second order by the products $(u')^i \, u_k''$. In this manner tensors of any order may be generated.

For $p = 1$, the transformation law

$$v = u \left| J\!\left(\begin{smallmatrix} x \\ y \end{smallmatrix}\right) \right|$$

defines a quantity called a **volume element**. The consistency of this definition follows from the relations

$$w = v \left| J\!\left(\begin{smallmatrix} z \\ y \end{smallmatrix}\right) \right| = u \left| J\!\left(\begin{smallmatrix} y \\ x \end{smallmatrix}\right) \right| \cdot \left| J\!\left(\begin{smallmatrix} z \\ y \end{smallmatrix}\right) \right| = u \left| J\!\left(\begin{smallmatrix} z \\ x \end{smallmatrix}\right) \right|.$$

Similarly the transformation law

$$v = u \, J\!\left(\begin{smallmatrix} y \\ x \end{smallmatrix}\right)$$

defines a quantity. More generally the laws

$$v = u \left[J\!\left(\begin{smallmatrix} y \\ x \end{smallmatrix}\right) \right]^r, \qquad r \text{ an integer,}$$

$$v = u \left| J\!\left(\begin{smallmatrix} y \\ x \end{smallmatrix}\right) \right|^r, \qquad r \text{ real,}$$

define quantities. In particular the quantity defined by setting $r = -1$ in the second relation is called a **density**. This terminology is motivated by the consideration of a mass distribution $\rho(x)$ in Euclidean space and the transformation law for its integral:

$$\int \rho(x) \, dx = \int \rho(x) \left| J\!\left(\begin{smallmatrix} x \\ y \end{smallmatrix}\right) \right| dy = \int \sigma(y) \, dy,$$

where

$$\sigma(y) = \rho(x) \left| J\!\left(\begin{smallmatrix} x \\ y \end{smallmatrix}\right) \right| = \rho(x) \left| J\!\left(\begin{smallmatrix} y \\ x \end{smallmatrix}\right) \right|^{-1}.$$

The relation

$$\sigma = \rho \, \text{sign} \left[J\!\left(\begin{smallmatrix} y \\ x \end{smallmatrix}\right) \right]$$

defines a quantity which we call an **orientation quantity**.

Let $\mathscr{S}$ be a manifold and $\mathscr{O}$ an open domain of $\mathscr{S}$. If to each point x of $\mathscr{O}$ we associate a quantity **u** of a definite type, then we say we have a **quantity field u** = **u**(x) defined in $\mathscr{O}$. If a quantity field **u**(x) is differentiable with respect to x, then this differentiability property is independent of the coordinate system. If **u** $= (u^1, \cdots, u^p)$ is a contravariant vector, then a change of coordinates $x \longrightarrow y$ implies that the change of the quantity **u** into **v** involves the first derivatives of the transformation of $x \longrightarrow y$. Hence the change of d**u** into d**v** involves the second derivatives of $x \longrightarrow y$. Thus to say that a contravariant vector field is once differentiable implies that the manifold must be smooth of order at least two. In general if **q** is a quantity field of order l defined in a manifold smooth of order k, with $k \geq l$, then the differentiability properties of **q** may exist only up to order $k - l$.

DEFINITION 14.1. A manifold $\mathscr{S}$, smooth of order $k \geq 1$, is **orientable** if there exists a continuous orientation quantity field defined over all of $\mathscr{S}$ which is nowhere zero.

Let $\mathscr{S}$ and $\mathscr{T}$ be n-dimensional manifolds, each smooth of order k, and suppose there is a one-one mapping of $\mathscr{S}$ onto $\mathscr{T}$. If x_0 is a point of $\mathscr{S}$ and y_0 a point of $\mathscr{T}$, suppose the mapping of a neighborhood of x_0 onto a neighborhood of y_0 is given by the functions $y^i = f_i(x)$, $i = 1, 2, \cdots, n$. If for each point of $\mathscr{S}$ these mapping functions are differentiable of order $l \leq k$, the mapping of $\mathscr{S}$ onto $\mathscr{T}$ is said to be **smooth of order l**.

Let **q** be a quantity of order $r \leq l$ defined at a point x_0 of $\mathscr{S}$. If **q** has components $u_1, u_2, \cdots, u_p$ in some coordinate system $x^1, x^2, \cdots, x^n$, then the transformation $y^i = f_i(x)$ will define the components of a quantity **q**$_1$ at the point y_0 of $\mathscr{T}$. For if $x \longrightarrow x'$ is a change of coordinates in a neighborhood of x_0, and $y \longrightarrow y'$ is the corresponding change of coordinates in the neighborhood of y_0, then **q** $\longrightarrow$ '**q**, and the fundamental homomorphism rule for quantities will also hold for **q**$_1 \longrightarrow$ (**q**$_1$)'.

If $\mathscr{U}$ is another n-dimensional manifold, smooth of order k, and if there is a one-one mapping of $\mathscr{T}$ onto $\mathscr{U}$, smooth of order l, then the quantity **q**$_1$ will be mapped into a quantity **q**$_2$. The mappings $\mathscr{S} \longrightarrow \mathscr{T} \longrightarrow \mathscr{U}$ induce a homomorphism, as the quantity **q**$_2$ obtained from **q** via **q**$_1$ is the same as the one obtained directly by the one-one mapping $\mathscr{S} \longrightarrow \mathscr{U}$ generated by composition of the mappings $\mathscr{S} \longrightarrow \mathscr{T}$ and $\mathscr{T} \longrightarrow \mathscr{U}$.

15. Generation of Quantities by Differentiation

Let **u**(x) be a quantity field of class C^1. The **complex** (**u**(x), ∂**u**$/\partial x$) is itself a quantity field. To see this, we suppose that **u** is a quantity of order l with the transformation functions given by

$$v_i = F_i\left(\mathbf{u}, \frac{\partial y}{\partial x}, \cdots, \frac{\partial^l y}{\partial x^l}\right),$$

where F_i is of class C^1. Then

$$\frac{\partial v_i}{\partial y^j} = \sum_{\sigma, \tau} \frac{\partial F_i}{\partial u_\sigma} \frac{\partial u_\sigma}{\partial x^\tau} \frac{\partial x^\tau}{\partial y^j} + \sum_{\rho, \sigma, \tau} \frac{\partial F_i}{\partial(\partial y^\rho/\partial x^\sigma)} \frac{\partial^2 y^\rho}{\partial x^\sigma \partial x^\tau} \frac{\partial x^\tau}{\partial y^j}$$

$$= F_{ij}\left(\mathbf{u}, \frac{\partial y}{\partial x}, \cdots, \frac{\partial^{l+1} y}{\partial x^{l+1}}\right).$$

Hence the complex $(\mathbf{u}(x), \partial \mathbf{u}/\partial x)$ maps to $(\mathbf{v}(y), \partial \mathbf{v}/\partial y)$ under the transformation $x \longrightarrow y$, and this complex is a quantity of order $l + 1$. In general the complexes thus formed will be of order one higher than the original quantities.

Example. The simplest quantity field is a **scalar field** u. This has the property that it is unchanged by any transformation of coordinates. That is, $v = u$ for any transformation $x \longrightarrow y$. Such a quantity is of order zero. We may apply the differentiation process and obtain a quantity $(u, \partial u/\partial x)$ of order 1. Since

$$du(x) = \sum_{i=1}^{n} \frac{\partial u}{\partial x^i} dx^i = dv(y) = \sum_{i=1}^{n} \frac{\partial v}{\partial y^i} dy^i,$$

we see that the quantity $\{\partial u/\partial x^i\}$, $i = 1, 2, \cdots, n$, is a covariant vector u_i, the **gradient** of u.

Let $u_i(x)$ be a covariant vector field. The quantity with components

$$\frac{\partial u_i}{\partial x^k} - \frac{\partial u_k}{\partial x^i} = u_{ik}$$

is a covariant tensor of order two. Here we see that the processes of differentiation and algebraic combination are employed to obtain a quantity of one higher order.

16. Commutator of Two Contravariant Vector Fields

Let $u^i(x)$, $v^i(x)$ be contravariant vector fields and let $f(x)$ be a scalar field. Since $\{\partial f/\partial x^i\}$ is a covariant vector, the quantity

$$g = \sum_{i=1}^{n} u^i(x) \frac{\partial f}{\partial x^i}$$

is a scalar. Similarly, since $\{\partial g/\partial x^i\}$ is a covariant vector, we again obtain a scalar by forming the combination

$$\sum_{j=1}^{n} \frac{\partial g}{\partial x^j} v^j = \sum_{i,j=1}^{n} u^i \frac{\partial^2 f}{\partial x^i \partial x^j} v^j + \sum_{i,j=1}^{n} \frac{\partial u^i}{\partial x^j} \frac{\partial f}{\partial x^i} v^j.$$

Interchanging the roles of u^i and v^i, we obtain the scalar

$$\sum_{i,j=1}^{n} v^i \frac{\partial^2 f}{\partial x^i \partial x^j} u^j + \sum_{i,j=1}^{n} \frac{\partial v^i}{\partial x^j} \frac{\partial f}{\partial x^i} u^j.$$

Subtracting the two scalars thus obtained, and noting that the first sums cancel, we obtain the scalar

$$\sum_{i=1}^{n} \left(\sum_{j=1}^{n} \frac{\partial u^i}{\partial x^j} v^j - \frac{\partial v^i}{\partial x^j} u^j \right) \frac{\partial f}{\partial x^i}.$$

Since f is arbitrary, we see that the quantities

$$w^i = \sum_{j=1}^{n} \left(\frac{\partial u^i}{\partial x^j} v^j - \frac{\partial v^i}{\partial x^j} u^j \right)$$

are the components of a contravariant vector $\mathbf{w}$. The vector $\mathbf{w}$ thus formed is called the **commutator** of the two contravariant vector fields $\mathbf{u}$ and $\mathbf{v}$. The process of forming the commutator is again one of differentiation and algebraic combination. In this case, however, we obtain a quantity of the same type as those we had originally. We also note that since the arbitrary scalar f had to be smooth of order two, the manifold also must be smooth of order at least two.

17. Hurwitz Integration on a Group Manifold

Let $\mathfrak{g}$ be a group manifold, smooth of order one. That is, the operations of composition, $\sigma \cdot \tau$, and formation of the inverse, τ^{-1}, are continuously differentiable. If α is a fixed element of $\mathfrak{g}$, then the transformations $\sigma \longrightarrow \alpha\sigma = \tau$ form a **left translation** group of transformations of $\mathfrak{g}$ to itself. These transformations are of class at least C^1.

THEOREM 17.1. *There is a density field ρ defined on all of $\mathfrak{g}$ which is invariant under the left translation group of $\mathfrak{g}$, and which is unique except for a constant factor.*

Proof. Let ϵ be the identity element of $\mathfrak{g}$. Fix ρ_0 as a quantity at ϵ. If α is the fixed element of the translation group, then ρ at α is defined by the rule for transforming quantities: In the transformation $\alpha\sigma = \tau$, we take for σ the identity ϵ and then $\tau = \alpha$. Hence we obtain a value of ρ at each point of g. We must still show that ρ is invariant under all transformations of the left translation group. Let $\alpha_1 \longrightarrow \alpha_2$, i.e., $\alpha\alpha_1 = \alpha_2$, be such a transformation. Since the group is simply transitive, there is a unique transformation that performs this. We consider the transformations

$\qquad$ 1. $\sigma \longrightarrow \alpha_1^{-1}\sigma = \sigma_1$

and

$\qquad$ 2. $\sigma_1 \longrightarrow \alpha_2\sigma_1 = \sigma_2$.

The value $\rho(\alpha_1)$ is transformed by (1) to $\rho_0(\epsilon)$, and the value $\rho_0(\epsilon)$ is transformed by (2) into $\rho(\alpha_2)$. Since the composition of these transformations is the unique transformation taking $\alpha_1 \longrightarrow \alpha_2$, we see that ρ is invariant. Since the value ρ_0 at ϵ was arbitrary, the density function is unique except for this constant factor.

Remark. In a similar way, we can obtain a density which is invariant under the right translation group. However, these densities may be different as is shown by the following example.

Example. The transformations

$$\begin{cases} x_1' = ax_1 + bx_2 \\ x_2' = cx_2 \end{cases} \qquad a > 0, \quad c > 0,$$

form a group characterized by the constants (a, b, c). Let (a, b, c) be the coordinates of the fixed element α in the group space:

$$\alpha = \begin{pmatrix} a & b \\ 0 & c \end{pmatrix}.$$

Then the composition $\alpha\sigma = \tau$ with

$$\sigma = \begin{pmatrix} x & y \\ 0 & z \end{pmatrix}, \qquad \tau = \begin{pmatrix} x' & y' \\ 0 & z' \end{pmatrix},$$

is given by

$$\begin{cases} x' = ax \\ y' = ay + bz \\ z' = cz. \end{cases}$$

The Jacobian of this transformation is

$$J\begin{pmatrix} x' & y' & z' \\ x & y & z \end{pmatrix} = \begin{vmatrix} a & 0 & 0 \\ 0 & a & b \\ 0 & 0 & c \end{vmatrix} = a^2c > 0.$$

Thus the left density field is $\rho_0(a, b, c) = 1/a^2c$; therefore in general, $\rho(x, y, z) = 1/x^2z$.

To obtain the right density field ρ^*, we have

$$\begin{pmatrix} x & y \\ 0 & z \end{pmatrix} \begin{pmatrix} a & b \\ 0 & c \end{pmatrix} = \begin{pmatrix} xa & xb + yc \\ 0 & zc \end{pmatrix},$$

or

$$\begin{cases} x' = ax \\ y' = bx + cy \\ z' = cz. \end{cases}$$

In this case the Jacobian is ac^2, hence $\rho^*(x, y, z) = 1/xz^2$.

THEOREM 17.2. *If $\mathfrak{g}$ is a compact group, the left and right densities are identical.*

Proof. Clearly, we have to show that the transformations $\sigma \longrightarrow \sigma' = \alpha\sigma\alpha^{-1}$ (all of which keep $\sigma = \epsilon$ fixed) take ρ_0 into itself for all α.

Let $s_1, s_2, \cdots, s_r$ be coordinate functions at ϵ, the identity, and let the transformation $\sigma \longrightarrow \sigma'$ be represented by $s_1, \cdots, s_r \longrightarrow s_1', \cdots, s_r'$. Then the density ρ_0 at ϵ transforms according to the law

$$\rho_0' = \rho_0 \left| J\begin{pmatrix} s' \\ s \end{pmatrix} \right|^{-1}, \qquad \text{evaluated at } s_i = 0.$$

The nth iterate of $\sigma \longrightarrow \sigma' = \alpha\sigma\alpha^{-1}$ is $\sigma \longrightarrow \sigma^{(n)} = \alpha^n\sigma\alpha^{-n}$. This sends $\epsilon \longrightarrow \epsilon$ and

$$\rho_0 \longrightarrow \rho_0 \left| J\begin{pmatrix} s \\ s' \end{pmatrix} \right|^{-n}, \qquad \text{evaluated at } s_i = 0.$$

Since $\mathfrak{g}$ is compact, the sequence $\{\alpha^n\}$ has a convergent subsequence $\alpha^{n_i} \longrightarrow \beta$. Since $\alpha^{n_i}\sigma\alpha^{-n_i} \longrightarrow \beta\sigma\beta^{-1}$, we see that the powers

$$\left| J\begin{pmatrix} s \\ s' \end{pmatrix} \right|^{-n_i}, \qquad \text{(evaluated at } s_i = 0\text{)}$$

must converge to a finite number different from zero. Hence

$$\left| J\begin{pmatrix} s \\ s' \end{pmatrix} \right| = 1.$$

It follows that $\rho_0' = \rho_0$.

LECTURE VII

18. Hurwitz Integration on a Group Manifold (continued)

Let $\mathscr{S}$ be a manifold, smooth of order ≥ 1 and ρ a continuous density field defined on $\mathscr{S}$. Let $\mathscr{T}$ be a closed subset of $\mathscr{S}$ and $f(p)$ a continuous function defined on $\mathscr{T}$. We make the following assumptions about $\mathscr{T}$:

1. $\mathscr{T}$ may be decomposed into a finite number of disjoint sets:

$$\mathscr{T} = \mathscr{T}_1 \cup \mathscr{T}_2 \cup \cdots \cup \mathscr{T}_k, \quad \mathscr{T}_i \cap \mathscr{T}_j = \varnothing, \quad i \neq j.$$

2. Each $\mathscr{T}_i$ is contained in an open set $\mathcal{O}_i$ in which a coordinate system can be introduced.

3. The coordinate system of $\mathcal{O}_i$, considered as a mapping x into Euclidean space, maps $\mathcal{O}_i$ onto a set $\Omega_i = x(\mathcal{O}_i)$. The corresponding mapping of $\mathscr{T}_i$ induced by this coordinate system will be a set $\tau_i = x(\mathscr{T}_i)$. It is supposed that each τ_i is Riemann measurable (has Jordan content.)

4. The closure of each $\mathscr{T}_i$ is compact in $\mathcal{O}_i$.

The function $f(p)$ on each $\mathscr{T}_i$ can be considered as a function $\phi(x^1, \cdots, x^n)$ of the coordinates $x^1, \cdots, x^n$ defined in Euclidean space. Similarly, the density function on $\mathscr{T}_i$ can be written $\rho(p) = \rho(x^1, \cdots, x^n)$. We form the Riemann integral

$$I_i = \int_{\tau_i} \phi(x^1, \cdots, x^n) \, \rho(x^1, \cdots, x^n) \, dx^1 \cdots dx^n.$$

We define the **integral** of $f(p)$ over the set $\mathscr{T}$ by the relation

$$\int_{\mathscr{T}} f(p) \, dp = \sum_{i=1}^{k} I_i.$$

It is necessary to show that the definition of integration is independent of the decomposition $\{\mathscr{T}_i\}$ and of the coordinate system selected. Let $\mathscr{T} = \mathscr{T}_1^* \cup \cdots \cup \mathscr{T}_l^*$ be another decomposition satisfying the same requirements as the first decomposition. Define the sets $\mathscr{T}_{ij} = \mathscr{T}_i \cap \mathscr{T}_j^*$, and form the expressions

$$I_{ij} = \int_{x(\mathscr{T}_{ij})} \phi(x)\, \rho(x)\, dx.$$

If $x \longrightarrow x^*$ is the transformation in $\mathscr{T}_i \cap \mathscr{T}_j^*$, then we have

$$\rho^*(x^*) = \rho(x) \left| J\!\left(\begin{matrix} x \\ x^* \end{matrix}\right) \right|,$$

and hence

$$I_{ij} = \int_{x^*(\mathscr{T}_{ij})} \phi(x^*)\, \rho^*(x^*)\, dx^*$$

by the usual rules for transforming Riemann integrals. Therefore

$$I_i = \sum_{j=1}^{l} I_{ij}, \qquad I_j^* = \sum_{i=1}^{k} I_{ij},$$

and

$$\sum_{i=1}^{k} I_i = \sum_{j=1}^{l} I_j^*.$$

19. Representation of Compact Groups

Let $\mathfrak{g}$ be a compact group manifold. We can cover $\mathfrak{g}$ by a finite number open sets $\mathcal{O}_i$, each with a coordinate system.

THEOREM 19.1. *Let $f(p)$ be continuous on $\mathfrak{g}$. Then*

$$\int_{\mathfrak{g}} f(p)\, dp$$

exists.

Proof. Let $\alpha \in \mathfrak{g}$. Then $\alpha \in \mathcal{O}_i$ for some i. Denote by Ω_i the set in Euclidean space which is the image of $\mathcal{O}_i$ under the coordinate mapping on $\mathcal{O}_i$. Let Σ_α be a closed sphere in Ω_i containing the map of the point α in its interior, and $\mathscr{S}_\alpha$ the inverse image of Σ_α in $\mathfrak{g}$. We can cover $\mathfrak{g}$ by a finite number of such sets $\mathscr{S}_1, \mathscr{S}_2, \cdots, \mathscr{S}_p$. Form the sets

$$\mathscr{T}_1 = \mathscr{S}_1, \quad \mathscr{T}_2 = \mathscr{S}_2 - \mathscr{S}_1, \quad \mathscr{T}_3 = \mathscr{S}_3 - (\mathscr{S}_1 \cup \mathscr{S}_2), \cdots.$$

The sets $\mathscr{T}_i$ satisfy the requirements for a decomposition of the manifold $\mathfrak{g}$; hence the integral exists.

Suppose $U(\alpha)$ is a finite representation of the compact group manifold $\mathfrak{g}$. That is, $U(\alpha)$ is an $n \times n$ matrix of functions $u_{ij}(\alpha)$, each of which is a continuous function of α. Furthermore $U(\alpha)U(\beta) = U(\alpha\beta)$ for all α, β in $\mathfrak{g}$.

THEOREM 19.2. *Each finite representation is equivalent to a unitary representation.*

Proof. The proof is essentially the same as the one for finite groups. (Cf. Theorems 5.4 and 5.6.) Let $H(\mathbf{x}, \mathbf{x})$ be a positive definite Hermitian form. That is,

$$H(\mathbf{x}, \mathbf{x}) = \sum h_{ij} x_i \overline{x_j} > 0 \qquad \text{for} \qquad \mathbf{x} \neq \mathbf{0}, \quad h_{ij} = \overline{h_{ji}}.$$

We consider transformations of the form

$$y_i = \sum_{j=1}^{n} u_{ij}(\alpha) \, x_j$$

and obtain

$$H(\mathbf{y}, \mathbf{y}) = H^{(\alpha)}(\mathbf{x}, \mathbf{x}).$$

We integrate the function $H^{(\alpha)}(\mathbf{x}, \mathbf{x})$ over $\mathfrak{g}$:

$$K(\mathbf{x}, \mathbf{x}) = \int_{\mathfrak{g}} H^{(\alpha)}(\mathbf{x}, \mathbf{x}) \, d\alpha.$$

(We assume $\int_{\mathfrak{g}} d\alpha = 1$, which is possible since the volume element has an arbitrary constant factor in its definition.) We now show, in analogy with the case of a finite group, that $K(\mathbf{x}, \mathbf{x})$ is invariant for all transformations of $\mathfrak{g}$. Let ω be such a transformation. Then

$$K^\omega(\mathbf{x}, \mathbf{x}) = \int_{\mathfrak{g}} H^{\alpha\omega}(\mathbf{x}, \mathbf{x}) \, d\alpha,$$

which follows from the fact that $(H^\alpha)^\beta = H^{\alpha\beta}$. We consider the right-hand translation $\alpha' = \alpha\omega$ and recall that the volume element is invariant under such a transformation; i.e., $d\alpha = d\alpha'$. Thus

$$K^\omega = \int_{\mathfrak{g}} H^{\alpha\omega}(\mathbf{x}, \mathbf{x}) \, d\alpha = \int_{\mathfrak{g}} H^{\alpha'}(\mathbf{x}, \mathbf{x}) \, d\alpha' = K(\mathbf{x}, \mathbf{x}),$$

since $\alpha \longleftrightarrow \alpha'$ is a homeomorphism. In a suitable coordinate system, K is the unit form, so the result follows.

DEFINITION 19.3. Let $f(\alpha)$, $g(\alpha)$ be continuous, complex-valued functions defined on $\mathfrak{g}$. The **inner product** of f and g is the number

$$f \cdot g = \int_{\mathfrak{g}} f(\sigma) \, \overline{g(\sigma)} \, d\sigma.$$

Also f and g are **orthogonal** if $f \cdot g = 0$.

THEOREM 19.4. *Let $U(\alpha)$, $V(\alpha)$ be irreducible, finite dimensional, inequivalent, representations of $\mathfrak{g}$. Then each component $u_{ij}(\alpha)$ of $U(\alpha)$ is orthogonal to each component $v_{kl}(\alpha)$ of $V(\alpha)$.*

Proof. The proof again follows the lines of that for finite groups. (Cf. Lemma 8.9.) Suppose $U(\alpha)$ is $m \times m$ and $V(\alpha)$ is $n \times n$. Let C be a mapping of the n-dimensional space to the m-dimensional space, and form the matrix A:

$$A = \int_{\mathfrak{g}} U(\alpha)\, C\, V^{-1}(\alpha)\, d\alpha,$$

where the integral of a matrix is the matrix formed by integrating each element. We shall show that $A = 0$. We multiply A by $U(\sigma)$ on the left and $V^{-1}(\sigma)$ on the right:

$$U(\sigma)\, A\, V^{-1}(\sigma) = \int_{\mathfrak{g}} U(\sigma)\, U(\alpha)\, C\, V^{-1}(\alpha)\, V^{-1}(\sigma)\, d\alpha$$

$$= \int_{\mathfrak{g}} U(\sigma\alpha)\, C\, V^{-1}(\sigma\alpha)\, d\alpha.$$

By the invariance of the volume element under left translations, we have $d\alpha = d(\sigma\alpha)$. Hence the equation yields $U(\sigma)\, A\, V(\sigma)^{-1} = A$.

We may now apply Lemma 8.9, which is independent of the finiteness of the groups. Thus, the inequivalence of U and V implies that $A = 0$. Therefore

$$\int_{\mathfrak{g}} U(\sigma)\, C\, V^{-1}(\sigma)\, d\sigma = 0$$

for all C. Take for C the matrix which has one in the ith row and jth column and zero elsewhere. Take U, V to be unitary, which is possible by Theorem 19.2. The result is

$$\int_{\mathfrak{g}} u_{ij}(\sigma)\, \overline{v_{kl}(\sigma)}\, d\sigma = 0 \qquad \text{for all} \quad i, j, k, l.$$

THEOREM 19.5. *If $U(\alpha)$ is a unitary representation of degree m, then*

$$\int_{\mathfrak{g}} u_{ij}(\sigma)\, \overline{u_{kl}(\sigma)}\, d\sigma = 0 \qquad \text{if} \quad (i, j) \neq (k, l),$$

$$\int_{\mathfrak{g}} u_{ij}(\sigma)\, \overline{u_{ij}(\sigma)}\, d\sigma = \frac{1}{m}\,.$$

The proof continues the direct analogy between finite and compact groups and repeats the argument leading to Lemma 9.3.

THEOREM 19.6. *The number of irreducible, inequivalent representations of a compact group is countable.*

Proof. For any two irreducible, inequivalent representations, the continuous functions u_{ij} and v_{kl} are orthogonal. The result follows because in the space C of continuous functions defined on a compact set, the number of mutually orthogonal functions is countable.

DEFINITION 19.7. If U_1, U_2, $\cdots$ is a sequence of irreducible, inequivalent representations such that each representation is similar to one of the U_i, then the sequence is said to form a **basis**.

We are leading to the following fundamental result.

THEOREM 19.8. *Let f be a continuous function on $\mathfrak{g}$. Then f may be approximated uniformly by finite linear combinations of the functions $u_{ij}{}^{(n)}(\alpha)$.*

A restatement of this theorem is the assertion that the functions $u_{ij}^{(n)}(\alpha)$ are uniformly complete in the space of continuous functions.

20. Existence of Representations

Let $K(\sigma, \tau)$ be a function of two variables defined on $\mathfrak{g}$, and consider the integral operator

$$g(\sigma) = \int_{\mathfrak{g}} K(\sigma, \tau) f(\tau) \, d\tau.$$

We seek conditions that make this integral operator invariant under right translation. That is,

$$\int_{\mathfrak{g}} K(\sigma, \tau) f(\tau\alpha) \, d\tau = g(\sigma\alpha).$$

Since

$$g(\sigma\alpha) = \int_{\mathfrak{g}} K(\sigma\alpha, \tau) f(\tau) \, d\tau = \int_{\mathfrak{g}} K(\sigma\alpha, \tau\alpha) f(\tau\alpha) \, d\tau,$$

we want

$$\int_{\mathfrak{g}} K(\sigma\alpha, \tau\alpha) f(\tau\alpha) \, d\tau = \int_{\mathfrak{g}} K(\sigma, \tau) f(\tau\alpha) \, d\tau.$$

This holds for all continuous f if and only if

$$K(\sigma\alpha, \tau\alpha) = K(\sigma, \tau).$$

Letting $\alpha = \tau^{-1}$, we find that $K(\sigma, \tau) = K(\sigma\tau^{-1}, \epsilon)$; hence $K(\sigma, \tau)$ is determined by a function k of one variable:

$$K(\sigma, \tau) = k(\sigma\tau^{-1}).$$

We recall several elementary theorems on integral equations.

THEOREM 20.1. *Let $K(\sigma, \tau)$ be a real-valued function, continuous in both variables and symmetric: $K(\sigma, \tau) = K(\tau, \sigma)$. Then there are non-identically vanishing solutions of* *the equation*

(†) $$\lambda \int_{\mathfrak{g}} K(\sigma, \tau) f(\tau) \, d\tau = f(\sigma)$$

for some values of the constant λ.

Each such value of λ which yields a non-vanishing solution is called an **eigenvalue**. A solution for this value is called an **eigenfunction**.

THEOREM 20.2. *There are at most a finite number of linearly independent eigenfunctions corresponding to a given eigenvalue.*

The number m of such functions is called the **multiplicity** of the eigenvalue. Obviously the eigenfunctions corresponding to an eigenvalue form a linear manifold.

Suppose a symmetric function $K(\sigma, \tau)$ has the form $K(\sigma, \tau) = k(\sigma\tau^{-1})$. Since the symmetry implies that $k(\sigma\tau^{-1}) = k(\tau\sigma^{-1})$, this is equivalent to stating that $k(\omega) = k(\omega^{-1})$ for all $\omega \in \mathfrak{g}$.

Let λ be an eigenvalue for such a kernel (as such functions $K(\sigma, \tau)$ are called) and let $f(\sigma)$ be an eigenfunction. Then for each α, the function $f(\sigma\alpha)$ is also an eigenfunction, because the integral equation (†) is invariant under right translations. Hence a right translation takes the space of eigenfunctions into itself, i.e., it induces a linear transformation.

The existence of a non-constant function $k(\omega)$ such that $k(\omega) = k(\omega^{-1})$ is easily demonstrated. Let $N(\epsilon)$ be a neighborhood of the identity, and let $l(\sigma)$ be any continuous function, positive in $N(\epsilon)$, zero on the boundary and exterior of $N(\epsilon)$. Define $k(\sigma) = l(\sigma) + l(\sigma^{-1})$; it is immediate that $k(\sigma)$ satisfies the required conditions.

LECTURE VIII

21. Representations of Compact Groups (continued)

We have seen that an integral operator

$$g(\sigma) = \int_{\mathfrak{g}} K(\sigma, \tau) f(\tau) \, d\tau$$

is invariant under the right-hand translations of the compact group $\mathfrak{g}$ if

$$K(\sigma, \tau) = k(\sigma\tau^{-1}),$$

where $k(\omega)$ is a continuous function on $\mathfrak{g}$.

We are particularly concerned with the case in which $K(\sigma, \tau)$ is a real symmetric kernel. This means that $k(\omega)$ is a continuous real-valued function and that $k(\omega) = k(\omega^{-1})$ so that $K(\tau, \sigma) = K(\sigma, \tau)$. We can always obtain a symmetric kernel from a given one by forming $\frac{1}{2}[k(\omega) + k(\omega^{-1})]$.

In general, a (not necessarily invariant) real symmetric kernel $K(\sigma, \tau)$ has eigenvalues, all real. If λ is one of these eigenvalues, then the number of linearly independent eigenfunctions which belong to λ is finite.

LEMMA 21.1. *If $K(\sigma, \tau)$ is a real symmetric kernel, invariant under right translations, if λ is an eigenvalue of K, and if $f(\sigma)$ is any eigenfunction belonging to λ, then for each α in $\mathfrak{g}$, the function $f(\sigma\alpha)$ is also an eigenfunction.*

Proof. The condition that K be invariant is that for any function $h(\tau)$,

$$\int_{\mathfrak{g}} K(\sigma, \tau) \, h(\tau\alpha) \, d\tau = \int_{\mathfrak{g}} K(\sigma\alpha, \tau) \, h(\tau) \, d\tau.$$

Since $f(\sigma)$ is an eigenfunction, we have

$$f(\sigma) = \lambda \int K(\sigma, \tau) f(\tau) \, d\tau.$$

Hence

$$f(\sigma\alpha) = \lambda \int K(\sigma\alpha, \tau) f(\tau)\, d\tau = \lambda \int_{\mathfrak{g}} K(\sigma, \tau) f(\tau\alpha)\, d\tau,$$

which means that $f(\sigma\alpha)$ is also an eigenfunction.

It follows from Lemma 21.1 that for each eigenvalue λ, the finite dimensional space of associated eigenfunctions is invariant under the right-hand translations of $\mathfrak{g}$. Thus we have a representation $\alpha \longrightarrow \|u_{ij}(\alpha)\|$ of degree m, where m is the multiplicity of λ, i.e., the number of linearly independent eigenfunctions.

We thus have a method of generating many representations of $\mathfrak{g}$; for we have a free choice of the function $k(\omega)$, and then each eigenvalue λ leads to a representation.

We previously proved that each representation of a compact group is similar to a unitary representation and is completely reducible. From Schur's lemma we obtained, exactly as we did for finite groups, the following results: Let $\|u_{ij}(\alpha)\|$, $\|v_{ij}(\alpha)\|$ be inequivalent, irreducible, unitary representations. Then

$$\int_{\mathfrak{g}} u_{ij}(\alpha)\, \overline{v_{kl}(\alpha)}\, d\alpha = 0,$$

that is, $u_{ij} \cdot v_{kl} = 0$. Here the metric was defined on the space of continuous functions on the group by

$$f \cdot g = \int_{\mathfrak{g}} f(\alpha)\, \overline{g(\alpha)}\, d\alpha.$$

It is understood that the volume of $\mathfrak{g}$ is normalized to 1. For the irreducible unitary representation $\|u_{ij}(\alpha)\|$ itself we proved

$$u_{ij} \cdot u_{kl} = 0 \qquad \text{if} \quad (i,j) \neq (k,l),$$

and

$$\int_{\mathfrak{g}} |u_{ij}|^2\, d\alpha = \frac{1}{m}.$$

It follows that the number of irreducible inequivalent (unitary) representations of a compact group is countable. For the contrary assumption leads to an uncountable orthonormal set of functions on $\mathfrak{g}$. This contradicts the separability of the compact space $\mathfrak{g}$.

Consequently there is a complete system $U_1(\alpha), U_2(\alpha), \cdots$ of representations of $\mathfrak{g}$. We expect this sequence to be infinite and the corresponding coordinate functions $u_{ij}^{(r)}(\alpha)$ to form a complete set of functions in the space of all continuous functions on $\mathfrak{g}$. By this we mean that there is no function other than zero which is orthogonal to all of the $u_{ij}^{(r)}(\alpha)$. Indeed, we shall prove that the set of functions $u_{ij}^{(r)}(\alpha)$ is uniformly complete.

THEOREM 21.2. *If $f(\alpha)$ is any continuous function on $\mathfrak{g}$ and $\eta > 0$, then there is a finite linear combination $g(\alpha)$ of the functions $u_{ij}^{(r)}(\alpha)$ such that*

$$|g(\alpha) - f(\alpha)| < \eta$$

for each $\alpha \in \mathfrak{g}$.

The proof of this result splits into several parts. First we quote a lemma from the theory of integral equations.

LEMMA 21.3. *Suppose $K(\sigma, \tau)$ is a symmetric kernel on $\mathfrak{g}$ and that $g(\sigma)$ is any function in the range of the corresponding integral operator, i.e.,*

$$g(\sigma) = \int_{\mathfrak{g}} K(\sigma, \tau)\, f(\tau)\, d\tau$$

for some continuous function $f(\tau)$ on $\mathfrak{g}$. Then $g(\sigma)$ can be uniformly approximated on $\mathfrak{g}$ by linear combinations of the eigenfunctions of $K(\sigma, \tau)$.

We do not prove the lemma, but point out that if $\phi_1(\sigma), \phi_2(\sigma), \cdots$ is the orthonormalized sequence of eigenfunctions of K, then the Fourier series

$$(g \cdot \phi_1)\phi_1(\sigma) + (g \cdot \phi_2)\phi_2(\sigma) + \cdots$$

converges uniformly on $\mathfrak{g}$ to $g(\sigma)$ when $g(\sigma)$ is in the range of K.

LEMMA 21.4. *If $K(\sigma, \tau) = k(\sigma\tau^{-1})$ is a symmetric right-invariant kernel, then each eigenfunction of K is a finite linear combination (with constant coefficients) of functions $u_{ij}^{(r)}$ derived from a representative system.*

Proof. Let λ be an eigenvalue of K, and let $f_1, \cdots, f_n$ be a basis for the space of corresponding eigenfunctions. We have seen that for each $\alpha \in \mathfrak{g}$, the right-hand translation by α sends each eigenfunction for λ into another and hence induces a representation $\|u_{ij}(\alpha)\|$ according to the equations

$$f_i(\sigma\alpha) = \sum_j u_{ij}(\alpha)\, f_j(\sigma).$$

We set $\sigma = \epsilon$, the identity, to obtain

$$f_i(\alpha) = \sum_j u_{ij}(\alpha)\, f_j(\epsilon) = \sum_j u_{ij}(\alpha)\, c_j.$$

Hence *each eigenfunction belonging to λ is a finite linear combination with constant coefficients of the functions $u_{ij}(\alpha)$.* We now obtain the lemma by simply decomposing the representation $\|u_{ij}(\alpha)\|$ into its irreducible parts.

It is clear that the eigenfunctions from a single kernel do not necessarily approximate every continuous function uniformly on $\mathfrak{g}$, since there may be only a finite number of eigenfunctions. However, we have considerable freedom in constructing kernels; now we shall use this freedom.

LEMMA 21.5. *Let $f(\sigma)$ be an arbitrary continuous function on $\mathfrak{g}$ and let $\eta > 0$. Then there exists a symmetric kernel $K(\sigma, \tau) = k(\sigma\tau^{-1})$ and a function $g(\sigma)$ in the range of this kernel such that*

$$|f(\sigma) - g(\sigma)| < \eta$$

for each σ in $\mathfrak{g}$.

Proof. We select a small neighborhood $\mathcal{N}$ of the identity ϵ of $\mathfrak{g}$ and select a function $k(\omega)$ which is positive on the neighborhood $\mathcal{N}$ and vanishes outside of $\mathcal{N}$. By averaging $k(\omega)$ and $k(\omega^{-1})$, we may assume $k(\omega) = k(\omega^{-1})$. Also we may multiply $k(\omega)$ by a positive constant so that

$$\int_{\mathfrak{g}} k(\omega)\, d\omega = 1.$$

We set $K(\sigma, \tau) = k(\sigma\tau^{-1})$ and

$$g(\sigma) = \int_{\mathfrak{g}} K(\sigma, \tau) f(\tau)\, d\tau = \int_{\mathfrak{g}} k(\sigma\tau^{-1}) f(\tau)\, d\tau.$$

To evaluate $g(\sigma)$ for a fixed value of σ, we note that $k(\sigma\tau^{-1}) = 0$ except when $\sigma\tau^{-1} \in \mathcal{N}^{-1}$ or $\tau \in \mathcal{N}\sigma$. By taking $\mathcal{N}$ sufficiently small in the first place, we can evidently insure that $|f(\tau) - f(\sigma)| < \eta$ for all σ and $\tau \in \mathcal{N}\sigma$. This implies that

$$|g(\sigma) - f(\sigma)| < \eta$$

for all $\sigma \in \mathfrak{g}$. Naturally we must use the uniform continuity of $f(\sigma)$ on the compact group $\mathfrak{g}$ to obtain this uniform approximation.

It is clear that the results of Lemmas 21.3, 21.4, and 21.5 combine to yield a proof of Theorem 21.2.

22. Characters

The character theory for compact groups is practically the same as for finite groups, with summation replaced by integration. The **character** of a representation $\|u_{ij}(\sigma)\|$ is the trace

$$\chi(\sigma) = \sum_i u_{ii}(\sigma).$$

This is a **primitive character** if the representation is irreducible. The characters $\chi(\sigma)$ are class functions on the group. If U_1, U_2 are inequivalent irreducible representations, then $\chi_1 \cdot \chi_2 = 0$ and $|\chi_i| = 1$. The following theorems are the analogues of results we had for finite groups.

THEOREM 22.1. *Each class function can be uniformly approximated by finite linear combinations of primitive characters.*

THEOREM 22.2. *The decomposition of a representation into irreducible components is essentially unique.*

THEOREM 22.3. *Two representations are similar if and only if they have the same character.*

23. Examples

1. *Rotation Group in the Plane*

This is the group of reals θ modulo 2π. Since it is a commutative group, each irreducible representation is one-dimensional. If m is an integer, the transformation

$$x' = e^{im\theta}x$$

yields a representation. These representations form a uniformly complete system. To prove this, we must show that each continuous function of period 2π is a uniform limit of trigonometric polynomials. But this is a well-known result of Weierstrass.

2. *Orthogonal Group in Three Dimensions with Determinant One*

For each $N = 0, 1, 2, \cdots$, we consider the totality of forms

$$f(x_1, x_2, x_3) = \sum_{i+j+k=N} a_{ijk} x_1{}^i x_2{}^j x_3{}^k,$$

a linear space of finite dimension. Under a rotation, each such form f goes into another form of the same degree; hence we have a representation of the rotation group. In this way we have constructed an infinite sequence of representations; in general these turn out to be reducible. However, for $N = 0$ the representation is obviously irreducible; for $N = 1$ it is also irreducible, because each linear form may be sent into a constant multiple of any other linear form.

For $N = 2$, we shall write the (quadratic) forms in the usual way, $f = \sum b_{ij} x_i x_j$ with $b_{ji} = b_{ij}$. The form $x_1{}^2 + x_2{}^2 + x_3{}^2$ spans a one-dimensional invariant subspace; hence the representation is reducible. We turn the representation into a unitary one by defining a metric on the space of quadratic forms:

$$f \cdot g = \int_\Omega f(x) \overline{g(x)} \, d\omega,$$

where $d\omega$ is the element of area on the unit sphere Ω: $x_1{}^2 + x_2{}^2 + x_3{}^2 = 1$. Having this, we see that the orthogonal complement of the one-dimensional invariant subspace noted above is the (five-dimensional) space of forms f satisfying

$$\int f \cdot (x_1{}^2 + x_2{}^2 + x_3{}^2)\, d\omega = 0,$$

i.e., $\int f\, d\omega = 0$. For $f = \sum b_{ij} x_i x_j$ this means $b_{11} + b_{22} + b_{33} = 0$. We shall prove that this space is irreducible.

First we note another way of describing the space. If Δ denotes the Laplace operator, then the condition $b_{11} + b_{22} + b_{33} = 0$ is equivalent to

$$\Delta f = \frac{\partial^2 f}{\partial x_1{}^2} + \frac{\partial^2 f}{\partial x_2{}^2} + \frac{\partial^2 f}{\partial x_3{}^2} = 0.$$

Of course the Laplacian of any function is invariant under rotation. This implies that for each N, the space $\mathscr{S}_N$ of all forms f of degree N satisfying $\Delta f = 0$ is an invariant subspace of the rotation group. We call $\mathscr{S}_N$ the space of **spherical harmonics** of degree N and shall prove the following theorem.

THEOREM 23.1. *Each of the spaces $\mathscr{S}_N$, $N = 1, 2, \cdots$, is irreducible. The corresponding representations of the rotation group on $\mathscr{S}_1, \mathscr{S}_2, \mathscr{S}_3, \cdots$ together form a complete system of representations of the rotation group.*

To prove this theorem, we need certain known results about spherical harmonics. First of all, each spherical harmonic of degree N is equal to ρ^N (where ρ is Euclidean distance from the origin), multiplied by a function of θ and ϕ alone, called a **surface harmonic** (Figure 2). A basis for the space of surface harmonics is easily expressed in terms of Legendre polynomials. We summarize the following information about Legendre polynomials. (See O. D. Kellogg, *Foundations of Potential Theory*, Springer-Verlag, Berlin, 1929, especially pp. 204–206; and E. T. Whittaker and G. N. Watson, *A Course of Modern Analysis*, 4th ed., Cambridge University Press, 1927, especially Chap. XV and pp. 391–395.)

The Legendre polynomial of degree N is given by

$$P_N(t) = \frac{1}{2^N (N!)} \left[(t^2 - 1)^N \right]^{(N)},$$

where (N) denotes N^{th} derivative. The functions $P_N(\cos \phi)$ are surface harmonics, in fact are the only surface harmonics invariant under rotation about the polar (x_3) axis. The **associated Legendre functions** are the functions

$$P_N{}^\nu(t) = (1 - t^2)^{\nu/2} [P_N(t)]^{(\nu)}, \qquad \nu = 1, 2, \cdots, N.$$

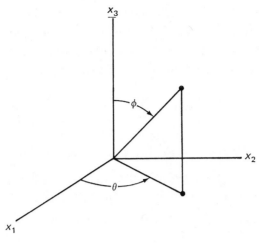

FIGURE 2

The functions

$$[\cos \nu\theta]P_N{}^\nu(\cos \phi), \qquad [\sin \nu\theta]P_N{}^\nu(\cos \phi),$$

are also surface harmonics, and these together with the $P_N(\cos \phi)$ form a basis of $\mathscr{S}_N$. Thus the most general surface harmonic $Y_N(\theta, \phi)$ of degree N is a linear combination

$$Y_N(\theta, \phi) = a_0 P_N(\cos \phi) + \sum_{\nu=1}^{N} (a_\nu \cos \nu\theta + b_\nu \sin \nu\theta)P_N{}^\nu(\cos \phi),$$

involving $(2N + 1)$ parameters.

We remark that $P_N(1) = 1$ and $P_N{}^\nu(1) = 0$ for $\nu = 1, 2, \cdots, N$.

We need some further results. Let P, Q be points on the unit sphere, ϕ_{PQ} the angular distance from P to Q. Then

$$P_N(\cos \phi_{PQ}),$$

when considered as a function of P with Q fixed, is a surface harmonic. Hence if $f(Q)$ is any continuous function on the sphere, then

$$\int_\Omega P_N(\cos \phi_{PQ}) f(Q) \, d\omega(Q)$$

is a surface harmonic. In particular it is true that if $Y_N(Q)$ is any surface harmonic of degree N, then

$$\frac{2N + 1}{4\pi} \int_\Omega P_N(\cos \phi_{PQ}) \, Y_N(Q) \, d\omega(Q) = Y_N(P).$$

Thus the functions $Y_N(Q)$ are eigenfunctions of $P_N(\cos \phi_{PQ})$. It can be shown that these are all of the eigenfunctions.

We now come to the proof that $\mathbf{S}_N$ is irreducible. We suppose that $\mathbf{L}$ is an invariant subspace of $\mathbf{S}_N$ and that there is a non-zero function $Y_N^0(\theta, \phi)$ in $\mathbf{L}$. By making a rotation we may assume Y_N^0 is different from 0 at the north pole; i.e., $Y_N^0(\theta, 0) \neq 0$. We expand:

$$Y_N^0(\theta, \phi) = a_0 P_N(\cos \phi) + \sum_{\nu=1}^{N} (a_\nu \cos \nu\theta + b_\nu \sin \nu\theta) P_N^\nu(\cos \phi).$$

Now

$$Y_N^0(\theta, 0) = a_0 P_N(1) = a_0, \qquad \text{hence} \quad a_0 \neq 0.$$

Since $\mathbf{L}$ is invariant under the rotation group, it follows that any rotation about the polar axis sends Y_N^0 into $\mathbf{L}$. Thus for each α

$$Y_N^0(\theta + \alpha, \phi) \in \mathbf{L},$$

and hence

$$\frac{1}{2\pi} \int_0^{2\pi} Y_N^0(\theta + \alpha, \phi) \, d\alpha \in \mathbf{L}.$$

We carry out the integration by noting that $P_N^\nu(\cos \phi)$ is independent of α, and that

$$\int_0^{2\pi} \cos \nu(\theta + \alpha) \, d\alpha = \int_0^{2\pi} \sin \nu(\theta + \alpha) \, d\alpha = 0.$$

Hence

$$\frac{1}{2\pi} \int_0^{2\pi} Y_N^0(\theta + \alpha, \phi) \, d\alpha = a_0 P_N(\cos \phi).$$

Since $a_0 \neq 0$, we conclude that $P_N(\cos \phi) \in \mathbf{L}$. By making a rotation, we further conclude that

$$P_N(\cos \phi_{PQ}) \in \mathbf{L} \qquad \text{for each } Q.$$

Finally, if Y_N is *any* surface harmonic, then

$$Y_N(P) = \frac{2N + 1}{4\pi} \int P_N(\cos \phi_{PQ}) \, Y_N(Q) \, d\omega(Q)$$

is in $\mathbf{L}$. Thus $\mathbf{L} = \mathbf{S}_N$, the space of all surface harmonics of degree N. In other words, $\mathbf{S}_N$ is irreducible.

We note that the proof may be completed another way. We have shown that each invariant subspace $\mathbf{L}$ of $\mathbf{S}_N$ contains the function $P_N(\cos \phi)$. If $\mathbf{L}$ is a proper subspace of $\mathbf{S}_N$, then its orthogonal complement is also invariant and hence also contains $P_N(\cos \phi)$, evidently impossible.

It remains to prove that the (irreducible) representations we have found, one for each $\mathbf{S}_N$, form a complete system. To do this, we shall compute the characters of these representations and prove that these characters form a complete system in the space of class functions on the rotation group.

First we must determine the conjugate classes. We see that any two axes of rotation may be sent by a rotation one to the other and that the rotation angle is preserved. Also by sending the axis of a rotation to the same axis with direction reversed, a rotation of angle ω is similar to a rotation of angle $-\omega$.

Summarizing: *Each equivalence class of rotations is uniquely determined by the rotation angle ω taken modulo 2π and modulo change of sign.* It follows that *the class functions are the even periodic continuous functions $f(\omega)$ of period 2π*:

$$f(-\omega) = f(\omega), \qquad f(\omega + 2\pi) = f(\omega).$$

A rotation of angle ω about the polar axis transforms the typical surface harmonic

$$Y_N = a_0 P_N(\cos\phi) + (a_1\cos\theta + b_1\sin\theta)P_N{}^1(\cos\phi) + \cdots$$

to

$$a_0 P_N(\cos\phi) + [(a_1\cos\omega + b_1\sin\omega)\cos\theta$$
$$+ (-a_1\sin\omega + b_1\cos\omega)\sin\theta]P_N{}^1(\cos\phi) + \cdots.$$

The trace of this transformation is clearly

$$1 + 2\cos\omega + 2\cos 2\omega + \cdots + 2\cos N\omega = \chi_N(\omega).$$

It is evident that the functions $\chi_N(\omega)$ form a basis for the space of even periodic functions since the functions 1, $\cos\omega$, $\cos 2\omega$, $\cdots$ are linear combinations of these functions.

It will be noted that we have obtained a complete system of representations of the rotation group without looking at the volume element of the group. We add a remark on the computation of this volume element. The typical rotation is a 3×3 matrix $A = \|a_{ij}\|$ satisfying

$$a_{i1}a_{j1} + a_{i2}a_{j2} + a_{i3}a_{j3} = \delta_{ij}.$$

There are nine elements a_{ij} and six independent equations, hence the group is three-dimensional. If X is an arbitrary orthogonal matrix, then $AX = Y$ describes the left-hand translation of X by A. Suppose however that X is an arbitrary matrix, i.e., an arbitrary point in nine-parameter space. Then the transformation $Y = AX$ has the property

$$\sum_{i,j} y_{ij}^2 = \sum_{i,j} x_{ij}^2.$$

In other words, distance is preserved so that this is a motion of nine-dimensional space which, of course, preserves area, volume, etc. The volume element of the rotation group is now obtained by computing the ordinary volume of the corresponding three-dimensional subspace of the nine-dimensional space.

3. *Unitary Group in Two Variables*

We consider the 2×2 complex matrices $U = \|u_{ij}\|$ satisfying $U\bar{U}' = I$. These are the matrices leaving the hermitian form $x_1\bar{x_1} + x_2\bar{x_2}$ invariant. Clearly

$$|\det U| = 1.$$

We shall restrict ourselves to the subgroup (which we call the **unitary group**) with $\det (U) = 1$.

We note that the transformations

$$U = \begin{bmatrix} \epsilon & 0 \\ 0 & 1/\epsilon \end{bmatrix}, \qquad \epsilon = e^{i\omega},$$

are unitary. Each unitary transformation is similar to one of these. If

$$V = \begin{bmatrix} \epsilon' & 0 \\ 0 & 1/\epsilon' \end{bmatrix}, \qquad |\epsilon'| = 1,$$

and $U \sim V$, then $\epsilon' = \epsilon$ or $\epsilon' = 1/\epsilon$. Thus the above matrix U uniquely describes a conjugate class provided ω is taken modulo 2π and modulo a change in sign.

We obtain representations for the group by considering for each n the space $\mathbf{T}_n$ of all binary forms of degree n,

$$a_0 x_1{}^n + a_1 x_1{}^{n-1}x_2 + \cdots + a_n x_2{}^n,$$

with complex coefficients. Each U transforms $\mathbf{T}_n$ into itself and hence $\mathbf{T}_n$ defines a representation of degree n. *These representations are irreducible and form a complete system.*

To prove this, let $\mathbf{L}$ be an invariant subspace of $\mathbf{T}_n$ such that there is a non-zero $f(x_1, x_2) = a_0 x_1{}^n + \cdots + a_n x_2{}^n \in \mathbf{L}$. Since any direction may be sent to $(1, 0)$, we may assume $f(1, 0) = a_0 \neq 0$. We apply

$$U = \begin{bmatrix} \epsilon & 0 \\ 0 & 1/\epsilon \end{bmatrix}, \qquad \epsilon = e^{i\omega},$$

to obtain

$$f\left(\epsilon x_1, \epsilon^{-1}x_2\right) = a_0 \epsilon^n x_1{}^n + a_1 \epsilon^{n-2} x_1{}^{n-1}x_2 + \cdots \in \mathbf{L}.$$

We divide by ϵ^n to obtain

$$a_0 x_1{}^n + a_1 \epsilon^{-2} x_1{}^{n-1}x_2 + \cdots \in \mathbf{L}.$$

Next we integrate on ω to obtain

$$\frac{1}{2\pi} \int_0^{2\pi} \epsilon^{-n} f\left(\epsilon x_1, x_2\right) d\omega \in \mathbf{L},$$

which yields simply $x_1{}^n \in \mathbf{L}$. The proof that $\mathbf{L} = \mathbf{T}_n$ is now completed as in the last example.

The completeness proof also proceeds as in the case of the rotation group. We must show that the characters of the representations on the spaces $\mathbf{T}_n$ form a complete system of class functions. But the trace of

$$f \longrightarrow f(\epsilon x_1, \epsilon^{-1} x_2)$$

is

$$\epsilon^n + \epsilon^{n-2} + \cdots + \epsilon^{-n} = 2 \cos n\omega + 2 \cos (n-2)\omega + \cdots$$

By taking suitable linear combinations of these characters for n even we obtain in turn 1, $\cos 2\omega$, $\cos 4\omega$, $\cdots$ and for n odd, $\cos \omega$, $\cos 3\omega$, $\cdots$, so altogether we obtain a complete system in the space of even periodic functions.

LECTURE IX

24. Lie Groups

Let $\mathfrak{g}$ be a continuous group represented by an r-dimensional manifold, smooth of order at least three. The group operations of composition and formation of an inverse are assumed to be continuously differentiable up to order three. We suppose furthermore that $\mathfrak{g}$ is **connected,** i.e., that any two elements σ and τ can be joined by an arc. If γ is such an arc connecting an element σ with the unit element ϵ, we may subdivide γ by selecting a finite set of elements on this curve beginning at ϵ and ending at σ: $\epsilon, \sigma_1, \sigma_2, \cdots, \sigma_n = \sigma$. Then the element σ can be written in the form

$$\sigma = (\sigma_n \sigma_{n-1}^{-1})(\sigma_{n-1} \sigma_{n-2}^{-1}) \cdots (\sigma_2 \sigma_1^{-1}) \sigma_1.$$

If the subdivision is sufficiently fine, all the factors $\sigma_i \sigma_{i-1}^{-1}$ will be close to the identity element. Thus each element σ of $\mathfrak{g}$ may be composed of elements taken from a given neighborhood $\mathcal{N}(\epsilon)$ of ϵ. This suggests a study of the group composition in the neighborhood of ϵ; since we assumed differentiability of the group composition, we may even consider an infinitesimal analysis of this composition (up to the order three).

Let us introduce a coordinate system in $\mathcal{N}(\epsilon)$, a neighborhood of ϵ. We normalize the coordinates so that ϵ has the coordinates $(0, 0, \cdots, 0)$. If σ, τ and $\sigma\tau = \omega$ are in $\mathcal{N}(\epsilon)$, we write in terms of coordinates

$$\sigma = (s^1, s^2, \cdots, s^r), \quad \tau = (t^1, t^2, \cdots, t^r), \quad \omega = (u^1, u^2, \cdots, u^r),$$

and we have

$$u^i = u^i(s, t), \qquad i = 1, 2, \cdots, r.$$

According to our assumptions, the u^i are three times continuously differentiable functions of the s^j and t^k in the neighborhood of $s^j = t^k = 0$. Since $\epsilon\tau = \tau$ and $\sigma\epsilon = \sigma$, we have at once

$$u^i(0, t) = t^i, \quad u^i(s, 0) = s^i, \qquad i = 1, 2, \cdots, r.$$

We expand the functions $u^i(s, t)$ in a neighborhood of the origin and obtain to the first order

$$u^i = s^i + t^i + \text{remainder term}.$$

This formula may be interpreted by stating that *up to the first order the group* $\mathfrak{g}$ *is identical with the translation group of dimension* r. Expanding one step further, we obtain

$$u^i = s^i + t^i + \sum_{\rho,\sigma=1}^{r} a_{\rho\sigma}^i s^\rho t^\sigma + \text{remainder term}.$$

Second-order terms containing only the s^i or only the t^i cannot appear because $u^i(0, t) = t^i$ and $u^i(s, 0) = s^i$. If instead of $\sigma\tau$ we compute $\tau\sigma$ and denote the coordinates of $\tau\sigma$ by v^i, we have

$$v^i = t^i + s^i + \sum a_{\rho\sigma}^i t^\rho s^\sigma + \text{remainder term}.$$

We form the differences $u^i - v^i = w^i$; the first-order terms drop out, and we find

$$w^i = \sum_{\rho,\sigma=1}^{r} (a_{\rho\sigma}^i - a_{\sigma\rho}^i) s^\rho t^\sigma + \text{remainder term}.$$

The second-order terms on the right-hand side of the last equation represent, one might say, a measure of the deviation from commutativity of the group composition if this composition is considered up to the second order. The quantities

$$a_{\rho\sigma}^i - a_{\sigma\rho}^i = c_{\rho\sigma}^i$$

are antisymmetric in the lower indices: $c_{\rho\sigma}^i = -c_{\sigma\rho}^i$. They were termed by Sophus Lie the **structure constants** of the group. It is clear that they depend on the coordinate system chosen.

The transformation character of the quantities $c_{\rho\sigma}^i$ can easily be obtained. If in the formulas for the w^i the remainder terms are dropped, so that

$$w^i = \sum_{\rho,\sigma=1}^{r} c_{\rho\sigma}^i s^\rho t^\sigma,$$

the corresponding relations in a new coordinate system,

$$\bar{w}^i = \sum_{\rho,\sigma=1}^{r} \bar{c}_{\rho\sigma}^i \bar{s}^\rho \bar{t}^\sigma,$$

are evidently obtained if the s^i, t^i, and w^i are transformed as components of contravariant vectors attached to ϵ. The $c^i_{\rho\sigma}$ are therefore the components of a mixed tensor at ϵ with one contravariant and two covariant indices. Thus the $c^i_{\rho\sigma}$ should better be called the components of the **structure tensor** of the group.

THEOREM 24.1. *It is possible to introduce a coordinate system in a neighborhood of ϵ so that*

$$a^i_{\rho\sigma} = \tfrac{1}{2}c^i_{\rho\sigma}.$$

Proof. We introduce a new coordinate system, $\bar{s}^i$ by

$$s^i = \bar{s}^i + \sum_{\rho,\sigma} \Gamma^i_{\rho\sigma}\bar{s}^\rho\bar{s}^\sigma,$$

where the $\Gamma^i_{\rho\sigma} = \Gamma^i_{\sigma\rho}$, but otherwise are still arbitrary. Since

$$u^i = s^i + t^i + \sum_{\rho,\sigma} a^i_{\rho\sigma}s^\rho s^\sigma + \text{remainder term,}$$

we have by substitution

$$\bar{u}^i + \sum_{\rho,\sigma} \Gamma^i_{\rho\sigma}\bar{u}^\rho\bar{u}^\sigma = \bar{s}^i + \sum_{\rho,\sigma} \Gamma^i_{\rho\sigma}\bar{s}^\rho\bar{s}^\sigma + \bar{t}^i + \sum_{\rho,\sigma} \Gamma^i_{\rho\sigma}\bar{t}^\rho\bar{t}^\sigma + \sum_{\rho,\sigma} a^i_{\rho\sigma}\bar{s}^\rho\bar{t}^\sigma + \cdots.$$

The left side is

$$\bar{u}^i + \sum \Gamma^i_{\rho\sigma}(\bar{s}^\rho + \bar{t}^\rho + \cdots)(\bar{s}^\sigma + \bar{t}^\sigma + \cdots) = \bar{u}^i + 2\sum \Gamma^i_{\rho\sigma}\bar{s}^\rho\bar{t}^\sigma + \cdots,$$

where we do not write the quadratic terms in $\bar{s}$ and in $\bar{t}$ as they automatically cancel. Hence we have

$$\bar{u}^i = \bar{s}^i + \bar{t}^i + \sum_{\rho,\sigma}(a^i_{\rho\sigma} - 2\Gamma^i_{\rho\sigma})\bar{s}^\rho\bar{t}^\sigma + \cdots.$$

If we now select the $\Gamma^i_{\rho\sigma}$ by the formulas

$$2\Gamma^i_{\rho\sigma} = \tfrac{1}{2}(a^i_{\rho\sigma} + a^i_{\sigma\rho}),$$

then

$$\bar{u}^i = \bar{s}^i + \bar{t}^i + \bar{a}^i_{\rho\sigma}\bar{s}^\rho\bar{t}^\sigma + \cdots,$$

where

$$\bar{a}^i_{\rho\sigma} = -\bar{a}^i_{\sigma\rho}.$$

Therefore in this coordinate system, we have $\bar{c}^i_{\rho\sigma} = 2\bar{a}^i_{\rho\sigma}$.

In obtaining the structure tensor of the group we only used the fact that $\sigma\epsilon = \sigma$ and $\epsilon\tau = \tau$. We now exploit the fact that the group operation is associative. In order to do this, however, we must make use of the assumption that the manifold is smooth of order at least three. If σ, τ, ω are any elements in $\mathcal{N}(\epsilon)$, we shall expand the products $(\sigma\tau)\omega$ and $\sigma(\tau\omega)$ in terms of the co-ordinate functions, and deduce what relations we can from the fact that these expansions must be identical.

For convenience we introduce the summation convention, which states that any repeated index in a given monomial represents the sum of all such mono-mials as the index goes from 1 to r.

We designate $\sigma\tau = \zeta$ with coordinates $\zeta = (z^1, z^2, \cdots, z^r)$. Then z^i has the expansion, to the third order,

$$z^i = s^i + t^i + a^i_{\rho\sigma} s^\rho t^\sigma + \tfrac{1}{2} f^i_{\rho\sigma\tau} s^\rho s^\sigma t^\tau + \tfrac{1}{2} g^i_{\rho\sigma\tau} s^\rho t^\sigma t^\tau + \cdots .$$

We note that all terms containing only products of the s^j or only products of the t^j are absent because $z^i(0, t) = t^i$ and $z^i(s, 0) = s^i$. Also, we have the symmetry relations

$$f^i_{\rho\sigma\tau} = f^i_{\sigma\rho\tau} \quad \text{and} \quad g^i_{\rho\sigma\tau} = g^i_{\rho\tau\sigma} .$$

We now form the expression $(\sigma\tau)\omega = \zeta\omega$ with coordinate functions u^i, and we write coordinates of $\omega = (w^1, w^2, \cdots, w^r)$. We have

$$u^i = s^i + t^i + a^i_{\rho\sigma} s^\rho t^\sigma + \tfrac{1}{2} f^i_{\rho\sigma\tau} s^\rho s^\sigma t^\tau + \tfrac{1}{2} g^i_{\rho\sigma\tau} s^\rho t^\sigma t^\tau$$

$$+ w^i + a^i_{\omega\tau}(s^\omega + t^\omega + a^\omega_{\rho\sigma} s^\rho t^\sigma + \cdots) w^\tau$$

$$+ \tfrac{1}{2} f^i_{\rho\sigma\tau}(s^\rho + t^\rho)(s^\sigma + t^\sigma) w^\tau$$

$$+ \tfrac{1}{2} g^i_{\rho\sigma\tau}(s^\rho + t^\rho)(w^\sigma)(w^\tau) + \cdots .$$

The mixed cubic terms of the above expression are

$$a^i_{\omega\tau} a^\omega_{\rho\sigma} s^\rho t^\sigma w^\tau + f^i_{\rho\sigma\tau} s^\rho t^\sigma w^\tau .$$

We next perform the same computation for the element $\sigma(\tau\omega)$ with coordinates $\bar{u}^i$, getting a sum of first-order terms, quadratic terms, and cubic terms as above. The first-order terms of w^i and $\bar{w}^i$ are clearly identical. By letting σ, τ, ω each be ϵ in turn, we see that all the quadratic terms in the expressions are also the same. The mixed cubic terms of $\bar{u}^i$ are given by

$$a^i_{\rho\omega} a^\omega_{\sigma\tau} s^\rho t^\sigma w^\tau + g^i_{\rho\sigma\tau} s^\rho t^\sigma w^\tau .$$

The associative law states that $u^i = \bar{u}^i$, hence

$$a^i_{\omega\tau} a^\omega_{\rho\sigma} + f^i_{\rho\sigma\tau} = a^i_{\rho\omega} a^\omega_{\sigma\tau} + g^i_{\rho\sigma\tau} .$$

We define

$$b^i_{\rho\sigma\tau} = a^i_{\omega\tau} a^\omega_{\rho\sigma} - a^i_{\rho\omega} a^\omega_{\sigma\tau} = g^i_{\rho\sigma\tau} - f^i_{\rho\sigma\tau} .$$

Since $f^i_{\rho\sigma\tau}$ is symmetric in the first two subscripts we have

$$b^i_{\rho\sigma\tau} - b^i_{\sigma\rho\tau} = g^i_{\rho\sigma\tau} - g^i_{\sigma\rho\tau} .$$

Furthermore since the $g^i_{\rho\sigma\tau}$ are symmetric in the last two subscripts, we have

$$g^i_{\rho\sigma\tau} - g^i_{\sigma\rho\tau} + g^i_{\sigma\tau\rho} - g^i_{\tau\sigma\rho} + g^i_{\tau\rho\sigma} - g^i_{\rho\tau\sigma} = 0,$$

or

$$b^i_{\rho\sigma\tau} - b^i_{\sigma\rho\tau} + b^i_{\sigma\tau\rho} - b^i_{\tau\sigma\rho} + b^i_{\tau\rho\sigma} - b^i_{\rho\tau\sigma} = 0.$$

By Theorem 24.1, we can introduce coordinates so that $a^i_{\rho\sigma} = \frac{1}{2}c^i_{\rho\sigma}$ is anti-symmetric in the subscripts. Then

$$4b^i_{\rho\sigma\tau} = c^i_{\omega\tau}c^\omega_{\rho\sigma} - c^i_{\rho\omega}c^\omega_{\sigma\tau}.$$

We introduce the notation

$$c^i_{\rho\sigma\tau} = c^i_{\omega\rho}c^\omega_{\sigma\tau},$$

and note that $c^i_{\rho\sigma\tau}$ is antisymmetric in the last two subscripts. Then

$$4b^i_{\rho\sigma\tau} = c^i_{\tau\rho\sigma} + c^i_{\rho\sigma\tau}.$$

Substituting this above, we find

(†) $$c^i_{\rho\sigma\tau} + c^i_{\sigma\tau\rho} + c^i_{\tau\rho\sigma} = 0.$$

Let $\mathbf{s} = (s^1, s^2, \cdots, s^r)$, $\mathbf{t} = (t^1, t^2, \cdots, t^r)$ be tangent vectors at ϵ. Then the vector $\mathbf{w} = (w^1, w^2, \cdots, w^r)$ defined by

$$w^i = \sum_{\rho,\sigma=1}^{r} c^i_{\rho\sigma} s^\rho t^\sigma$$

is also a tangent vector. (See the paragraph preceding Theorem 24.1.) We say $\mathbf{w}$ is obtained by **commutator multiplication** (**composition**) of $\mathbf{s}$ and $\mathbf{t}$ and write

$$\mathbf{w} = [\mathbf{s}\,\mathbf{t}].$$

With this definition the relation (†) may be written

(*) $$[[\mathbf{s}\,\mathbf{t}]\,\mathbf{u}] + [[\mathbf{t}\,\mathbf{u}]\,\mathbf{s}] + [[\mathbf{u}\,\mathbf{s}]\,\mathbf{t}] = 0.$$

Commutator multiplication is *anti-commutative*, since the relation $c^i_{\rho\pi} = -c^i_{\sigma\rho}$ implies that

$$[\mathbf{s}\,\mathbf{t}] = -[\mathbf{t}\,\mathbf{s}].$$

If we consider the n-dimensional tangent vector space at ϵ, the elements $\mathbf{s}, \mathbf{t}, \cdots$ with the multiplication described above form an algebra, termed a **Lie algebra**. This algebra is anti-commutative, and the cyclic identity (*) replaces the usual associativity rule. We call this algebra the **infinitesimal group** of $\mathfrak{g}$, and the elements of this algebra, the tangent vectors of $\mathfrak{g}$ at ϵ, we call the **infinitesimal elements** of $\mathfrak{g}$.

25. Infinitesimal Transformations on a Manifold

We shall now introduce the important concept of an infinitesimal transformation on a manifold $\mathscr{S}$. We assume that $\mathscr{S}$ is smooth of order at least two, and consider a one-parameter family of one-one transformations of $\mathscr{S}$ onto itself:

$$p_0 \longrightarrow p = \mathbf{f}(p_0, t) \qquad (-t_0 < t < t_0).$$

It is convenient to use physical terminology and to interpret t as a time variable and $\mathbf{f}$ as giving the position p of a particle, characterized by p_0, at the time t. The simplest choice of the Lagrangian label p_0 is the position of the moving particle at a definite time, say $t = 0$, in which case $\mathbf{f}(p_0, 0) = p_0$.

Assume now that $p = \mathbf{f}(p_0, t)$ is continuous and has a continuous derivative $\partial \mathbf{f}(p_0, t)/\partial t$. In the language of fluid motion, this means that each particle moves with a definite velocity. We look on the derivative as a function of p and t instead of p_0 and t, and obtain

$$\frac{\partial \mathbf{f}(p_0, t)}{\partial t} = \mathbf{U}(p, t),$$

a contravariant vector field spread over $\mathscr{S}$, depending in general on the parameter t. We call it the **infinitesimal transformation** derived from the one-parameter family of transformations $p = \mathbf{f}(p_0, t)$. It gives the velocity vector $\mathbf{U}(p, t)$ of the particle which at time t occupies the position p.

The problem of reconstructing the original one-parameter family of transformations $p = \mathbf{f}(p_0, t)$ from the derived infinitesimal transformation is evidently equivalent to solving the ordinary differential equation

$$\frac{dp}{dt} = \mathbf{U}(p, t).$$

Written in local coordinates, this means solving the system

$$\frac{dx^i}{dt} = u^i(x^1, x^2, \cdots, x^n, t) \qquad (i = 1, 2, \cdots, n),$$

where the u^i represent the components of $\mathbf{U}$ at the point p with coordinates x^i at time t. If an assumption is introduced which insures existence and uniqueness of a solution with given initial conditions, e.g., continuous differentiability of $\mathbf{U}(p, t)$ in p, we see that there exists one and only one $\mathbf{f}(p_0, t)$ satisfying our requirement and the initial condition $\mathbf{f}(p_0, 0) = p_0$.

We shall introduce the concept of the **commutator** of two infinitesimal transformations. We start with two one-parameter families of transformations on $\mathscr{S}$, say $\sigma(s)$ and $\tau(t)$, such that $\sigma(0)$ and $\tau(0)$ represent the identity transformation. We define the mappings

$$\omega_{st} = \sigma_s \tau_t \qquad \omega_{st}^* = \tau_t \sigma_s.$$

We are interested in how much ω_{st} and ω_{st}^* deviate from each other for small values of the parameters s and t. The following consideration facilitates the computation of this deviation. Let $F(p)$ be an arbitrary function on $\mathscr{S}$ of class C'. If $p = \sigma_s p_0$ is the image of p_0 under σ_s, the substitution of $p = \sigma_s p_0$ into $F(p)$ gives a function

$$\Phi(p_0, s) = F(\sigma_s p_0).$$

Let $x^1, x^2, \cdots, x^n$ be a coordinate system around p_0. Then

$$\mathbf{U}F = \frac{\partial \Phi}{\partial s}$$

$$= \sum_{\rho=1}^{n} \frac{\partial F}{\partial x^\rho} \frac{dx^\rho}{ds} = \sum \frac{\partial F}{\partial x^\rho} u^\rho$$

represents a linear homogeneous differential operator $\mathbf{U}$ of first order. Its coefficients u^i are the components of the infinitesimal transformation derived from σ_s at the parameter value s. We form

$$F(\omega_{st} p_0) = \psi(p_0, s, t),$$
$$F(\omega_{st}^* p_0) = \psi^*(p_0, s, t).$$

The deviation of ψ and ψ^* for small values of s and t and for arbitrary $F(p)$ is a measure of the deviation from commutativity of σ_s and τ_t.

We now make the assumptions that F is of class C^2, and that the infinitesimal transformations $\mathbf{U}$ and $\mathbf{V}$, derived from σ_s and τ_t respectively, are of class C^1. Since ω_{st} and ω_{st}^* coincide if $s = 0$ or $t = 0$, we obtain after some calculation the Taylor expansion for $\psi - \psi^*$ up to the second order at $s = t = 0$:

$$\psi(p_0, s, t) - \psi^*(p_0, s, t) = (\mathbf{W}F)st + \text{remainder term.}$$

We see that $\mathbf{W}F$ is a linear homogeneous differential operator of first order,

$$\mathbf{W}F = (\mathbf{V}\mathbf{U} - \mathbf{U}\mathbf{V})F = \sum w^i \frac{\partial F}{\partial x^i},$$

with components

$$w^i = \sum_\rho \frac{\partial u^i}{\partial x^\rho} v^\rho - \frac{\partial v^i}{\partial x^\rho} u^\rho \qquad \text{at} \quad s = t = 0.$$

The quantity $\mathbf{W}F$ is called the **commutator** of the infinitesimal transformations $\mathbf{U}$ and $\mathbf{V}$ at $s = t = 0$.

LECTURE X

26. Infinitesimal Transformations of a Group

In what follows, we shall suppose that the functions concerned possess sufficient differentiability to make all operations valid. Let $\mathfrak{g}$ be a group, smooth of class three, which is realized as a group $\mathfrak{G}$ of transformations of a space $\mathscr{S}$. Thus to each σ in $\mathfrak{g}$ corresponds a transformation $T(\sigma)$. If $p_0 \in \mathscr{S}$, we write

$$p = T(\sigma)p_0 = \mathbf{f}(p_0, \sigma).$$

We have $T(\alpha\beta) = T(\alpha) T(\beta)$. We assume this realization is **faithful:** $T(\alpha) = I$ only when $\alpha = \epsilon$.

Let $\sigma(t)$ be a differentiable curve in $\mathfrak{g}$. Then $T[\sigma(t)]$ is a one-parameter family of transformations on $\mathscr{S}$; we compute the infinitesimal transformation. At each point p of $\mathscr{S}$, it is the vector

$$\left. \frac{dp}{dt} \right|_{t_0},$$

where $p = \mathbf{f}(p_0, \sigma(t))$. The totality of infinitesimal transformations obtained from all possible curves $\sigma(t)$ is called the set of **infinitesimal transformations** of the group $\mathfrak{g}$.

We shall first show that there are not many infinitesimal transformations of $\mathfrak{g}$. To simplify matters we shall take $t_0 = 0$, which is always possible by a simple translation of the time axis. Next, we may assume the curve starts at ϵ, because the family

$$T[\sigma(t)\sigma^{-1}(0)] = T[\sigma(t)] \{T[\sigma(0)]\}^{-1}$$

has the same infinitesimal transformation at $t = 0$ as does $T[\sigma(t)]$. For convenience we set $\tau(t) = \sigma(t)\sigma^{-1}(0)$, so that $\tau(0) = \epsilon$. We let s^j be parameters at ϵ in $\mathfrak{g}$ and x^i parameters at p in $\mathcal{S}$. Thus the transformation

$$p = T(\tau)p_0$$

is represented by

$$x^i = f^i(x_0{}^1, \cdots, x_0{}^n; s^1, \cdots, s^r).$$

The path $\tau = \tau(t)$ with $\tau(0) = \epsilon$ is represented by

$$s^i = s^i(t),$$

and we may assume $s^i(0) = 0$. We compute the velocity field:

$$\left(\frac{dx^i}{dt}\right)_{t=0} = \left(\frac{\partial f^i}{\partial s^\rho}\right)_{s=0} \left(\frac{ds^\rho}{dt}\right)_{t=0}.$$

Since $f^i(x_0, 0) = x_0{}^i$, it is clear that

$$\left(\frac{\partial f^i}{\partial s^\rho}\right)_{s=0} = u_\rho^i(x_0);$$

hence

$$u^i = \left(\frac{dx^i}{dt}\right)_{t=0} = u_\rho^i(x) \left(\frac{ds^\rho}{dt}\right)_{t=0}$$

represents the infinitesimal transformation at the point p with coordinates x^i. It is clear that the only feature of the path which enters into its determination is the tangent vector

$$\left(\frac{ds^\rho}{dt}\right)_{t=0}$$

to the path at the origin ϵ. Thus we have a linear mapping from vectors at ϵ on $\mathfrak{g}$ to infinitesimal transformations. Since the space of vectors at ϵ has dimension r, it follows that *there are at most r linearly independent infinitesimal transformations of $\mathfrak{g}$ on $\mathcal{S}$.*

Let us take the basis

$$(1, 0, \cdots, 0)', \ (0, 1, 0, \cdots, 0)', \cdots, (0, 0, \cdots, 1)'$$

of the space of tangent vectors to $\mathfrak{g}$ at ϵ. Corresponding to these vectors are r infinitesimal transformations of $\mathfrak{g}$ on $\mathcal{S}$:

$$\mathbf{U}_1 = (u_1{}^1, \cdots, u_1{}^n)'$$
$$\mathbf{U}_2 = (u_2{}^1, \cdots, u_2{}^n)'$$
$$\cdot \quad \cdot \quad \cdot \quad \cdot \quad \cdot \quad \cdot \quad \cdot \quad \cdot$$
$$\mathbf{U}_r = (u_r{}^1, \cdots, u_r{}^n)'.$$

They necessarily span the space of all infinitesimal transformations of $\mathfrak{g}$. After looking at some examples, we shall prove the following fundamental result.

THEOREM 26.1. *The infinitesimal transformations* $\mathbf{U}_1, \cdots, \mathbf{U}_r$ *of a faithful representation are linearly independent.*

This means that no linear combination $\sum a^i \mathbf{U}_i$ is identically 0 for all p. As the examples will show, a linear combination may vanish at certain points.

27. Examples

Example 1.

$$x = x_0 + s$$

on the real line. Here $\partial x/\partial s = 1$, hence $\mathbf{U}(x) = \mathbf{I}$, an infinitesimal translation on the real axis.

Example 2.

$$x = ax_0 + b, \qquad a > 0.$$

The unit element ϵ is given by $a = 1, b = 0$. We have

$$\left(\frac{\partial x}{\partial a}\right)_\epsilon = (x_0)_{a=1, b=0} = x, \qquad \left(\frac{\partial x}{\partial b}\right)_\epsilon = 1.$$

Hence $u_1^1 = x$, $u_2^1 = 1$, and the general infinitesimal transformation is $\mathbf{U}(x) = a_1 x + a_2$.

Example 3.

$$x = \frac{ax_0 + b}{cx_0 + 1}, \qquad \begin{vmatrix} a & b \\ c & 1 \end{vmatrix} > 0.$$

The identity transformation is given by

$$\begin{pmatrix} a & b \\ c & 1 \end{pmatrix} = \begin{pmatrix} 1 & 0 \\ 0 & 1 \end{pmatrix}.$$

At the identity, we have

$$\frac{\partial x}{\partial a} = x, \quad \frac{\partial x}{\partial b} = 1, \quad \frac{\partial x}{\partial c} = -x^2,$$

hence the general infinitesimal transformation of the group is

$$a_0 + a_1 x + a_2 x^2.$$

Example 4. Orthogonal group:

$$x_i = \sum a_{i\rho} x_\rho^0, \qquad A = \|a_{i\rho}\|, \quad AA' = I.$$

We think of $A = A(t)$, a one-parameter family of orthogonal transformations with $A(0) = I$. Then

$$\frac{dx_i}{dt} = \sum \frac{da_{i\rho}}{dt} (x_\rho^0)_{t=0} = \frac{da_{i\rho}}{dt} x_\rho,$$

or, in matrix notation,

$$\frac{dX}{dt} = \frac{dA}{dt} X.$$

However,

$$\frac{dA}{dt} A' + A \frac{dA'}{dt} = 0.$$

At $t = 0$ we have $A = A' = I$, hence

$$\frac{dA}{dt} + \left(\frac{dA}{dt}\right)' = 0 \quad \text{at} \quad t = 0.$$

Thus

$$B = \left(\frac{dA}{dt}\right)_{t=0}$$

is skew-symmetric and represents the general infinitesimal transformation of the group.

Conversely, if B is a (constant) skew-symmetric matrix then

$$A(t) = e^{Bt}$$

is a one-parameter family of orthogonal transformations such that

$$A(0) = I, \quad \left(\frac{dA}{dt}\right)_{t=0} = B.$$

Example 5. Unitary group:

$$X = AX_0, \quad A\bar{A}' = I.$$

The analysis is similar to that given for the orthogonal group. We must find all matrices

$$B = \left(\frac{dA}{dt}\right)_{t=0}$$

corresponding to one-parameter families $A(t)$ of unitary matrices satisfying $A(0) = I$. We evidently obtain

$$B + \bar{B}' = 0.$$

Hence if we set $H = iB$, then H is Hermitian. Conversely, each unitary matrix arises in this way.

Example 6. General linear group:

$$X = AX_0, \quad |A| \neq 0.$$

Here we consider $A = A(t)$ with $A(0) = I$. The result states that

$$B = \left(\frac{dA}{dt}\right)_{t=0}$$

is an arbitrary matrix and

$$X \longrightarrow BX$$

is the general infinitesimal transformation.

28. Geometry on the Group Space

We return to Theorem 26.1 in which we promised to show that the infinitesimal transformations

$$\mathbf{U}_1, \cdots, \mathbf{U}_r$$

are linearly independent. To do this we introduce the concept of **parallel vectors** on the group space $\mathfrak{g}$.

Let σ_1, σ_2 be points of $\mathfrak{g}$ and consider the right translates $\sigma_1\alpha$, $\sigma_2\alpha$. By analogy with Euclidean space we shall call an ordered pair (σ, τ) of elements of $\mathfrak{g}$ a **vector** (directed segment from σ to τ). We shall call the vectors $(\sigma_1, \sigma_1\alpha)$ and $(\sigma_2, \sigma_2\alpha)$ **right-hand parallel**. If we set $\tau_1 = \sigma_1\alpha$, $\tau_2 = \sigma_2\alpha$, then the equation

$$\sigma_2\sigma_1^{-1} = \tau_2\tau_1^{-1}$$

characterizes the parallelism of the vectors (σ_1, τ_1) and (σ_2, τ_2). Similarly we introduce left parallelism of vectors: $(\sigma_1, \alpha\sigma_1)$ and $(\sigma_2, \alpha\sigma_2)$ are **left-hand parallel**. We note the result: (σ_1, τ_1) and (σ_2, τ_2) are right-hand parallel if and only if (σ_1, σ_2) and (τ_1, τ_2) are left-hand parallel.

Next we extend this concept to infinitesimal vectors. Let $\sigma(t)$ be a curve and let $\tau(t) = \sigma(t)\alpha$ be the right-hand translate of this curve by α: thus $\tau(0) = \sigma(0)\alpha = \sigma\alpha = \tau$. The tangent vector at σ is $(d\sigma/dt)$ and that at τ is $(d\tau/dt)$. They are related by

$$\frac{d\tau}{dt} = \frac{d\sigma}{dt}\alpha.$$

We shall say that the vectors $(d\sigma/dt)$ at σ and $(d\tau/dt)$ at τ are **left-hand parallel** to each other. If σ and τ are given in advance, there is a unique α such that $\sigma\alpha = \tau$; hence we have a one-one map of the infinitesimal transformations at σ to those at τ. Evidently this mapping is a linear one.

We note that if $\omega = \tau\beta = \sigma\alpha\beta$, then the composition of these linear transformations, first from σ to τ, then from τ to ω, is the same as the linear transformation from σ to ω directly. In other words, parallelism is transitive. We similarly prove the symmetry and reflexivity of the parallelism relation.

Also, from the simple transitivity of the right-hand translation group, we deduce that each vector at α is parallel to exactly one vector at β.

We noted above that when $T(\sigma)$ is a faithful realization of $\mathfrak{g}$ as a transformation group on a space $\mathscr{S}$, then $T[\sigma(t)]$ and $T[\sigma(t)\alpha]$ yield the same infinitesimal transformation on $\mathscr{S}$. But

$$\frac{d\tau}{dt} = \frac{d\sigma}{dt}\,\alpha \quad \text{and} \quad \frac{d\sigma}{dt}$$

are parallel. Hence: *parallel tangent vectors on $\mathfrak{g}$ determine the same infinitesimal transformation on $\mathscr{S}$.*

With this last result, we may proceed to the proof that $\mathbf{U}_1, \cdots, \mathbf{U}_r$ are linearly independent. If not, there are constants a^i, not all zero, such that

$$a^1\mathbf{U}_1 + \cdots + a^r\mathbf{U}_r = 0.$$

We consider $(a^1, \cdots, a^r)$ as a vector at ϵ and place a parallel vector $(b^1, \cdots, b^n)$ at each point σ of $\mathfrak{g}$, forming a vector field $a(\sigma)$ on $\mathfrak{g}$. By solving the ordinary differential equation

$$\frac{d\sigma}{dt} = a(\sigma)$$

with initial condition $\sigma(0) = \epsilon$, we obtain a curve $\sigma(t)$ on $\mathfrak{g}$ whose tangent vector at any point is $a(\sigma)$. It follows that the infinitesimal transformation on $\mathscr{S}$ arising from this curve is $\sum a^i\mathbf{U}_i = 0$. However each point on the curve gives rise to the same transformation $T(\sigma)$ of $\mathscr{S}$, since the tangents at the various points of the curve are all parallel. Since $dp/dt = 0$, it follows that $T(\sigma)$ is constant along the curve. This contradicts the assumption that the representation is faithful.

LECTURE XI

29. Parallelism

Let $\mathfrak{g}$ be an r-dimensional group manifold. We introduce the coordinate system $s^1, s^2, \cdots, s^r$ in a neighborhood of the identity ϵ, and we let $t^1, t^2, \cdots, t^r$ be a coordinate system in a neighborhood of the point $\alpha \in \mathfrak{g}$. Then the transformation $\sigma \longrightarrow \tau = \sigma\alpha$ takes any vector at $\sigma = \epsilon$ into a vector at α. More specifically, if we introduce the functions

$$\psi_\rho^i(\alpha) = \left(\frac{\partial s^i}{\partial t^\rho}\right)_{\tau=\alpha},$$

then we have

(*) $$ds^i = \sum_{\rho=1}^{r} \psi_\rho^i(\alpha)\, dt^\rho.$$

Since the transformation is one-one, the matrix (ψ_ρ^i) is non-singular. We write briefly

$$ds = \psi\, dt, \qquad \text{where} \quad \psi = \| \psi_\rho^i(\alpha) \|,$$

with i the row index, ρ the column index, and

$$ds = \begin{pmatrix} ds^1 \\ \cdot \\ \cdot \\ \cdot \\ ds^r \end{pmatrix}, \qquad dt = \begin{pmatrix} dt^1 \\ \cdot \\ \cdot \\ \cdot \\ dt^r \end{pmatrix},$$

and we say that a vector ds at ϵ is **left-parallel** to a vector dt at α if the components of the vectors are related according to formula (*).

If α and β are two elements of $\mathfrak{g}$, and dt is a vector at α and du a vector at β, then we say the two vectors are **left-parallel** if the right translation transforming α into β transforms dt into du. This means evidently that

$$\psi(\alpha) \, dt = \psi(\beta) \, du, \qquad du = \psi^{-1}(\beta) \, \psi(\alpha) \, dt.$$

If we change coordinates at ϵ then the matrix ψ becomes a matrix $\psi_0\psi$, where ψ_0 is a constant non-singular matrix. It is also formally obvious that this does not affect the parallelism of vectors, since

$$du = [\psi_0\psi(\beta)]^{-1}[\psi_0\psi(\alpha)] \, dt = \psi^{-1}(\beta) \, \psi(\alpha) \, dt.$$

Thus ψ is determined only up to a constant non-singular matrix factor on the left. For each fixed i, the ψ_ρ^i, $\rho = 1, 2, \cdots, r$, are the components of a covariant vector field. Hence a change of coordinates at α means a change according to the laws of transformation of covariant vectors.

THEOREM 29.1. *The vector field formed by attaching to each element α of $\mathfrak{g}$ the vector left-parallel to a fixed vector at ϵ is an infinitesimal transformation of the left-hand translation group.*

Proof. Consider a curve $\sigma = \sigma(p)$, $-p_0 < p < p_0$, starting at $\sigma(0) = \epsilon$. To the curve there corresponds a one-parameter family of left-hand translations transforming an arbitrary element α into $\tau = \sigma(p)\alpha$. In order to obtain the associated infinitesimal transformation for $p = 0$, we have to compute, for fixed α, the vector $(d\tau/dp)_{p=0}$ attached to the point α. But from the definition of left-parallelism of vectors of $\mathfrak{g}$, we see that all vectors thus obtained are left-parallel to the vector $(d\sigma/dp)_{p=0}$ at ϵ. Since the latter vector may be arbitrarily chosen, the result follows.

Now let u^i be a vector at ϵ and v^i the left-parallel vector at σ, so that

$$u^i = \sum_\rho \psi_\rho^i(\sigma)v^\rho.$$

We define $\phi = \psi^{-1}$, which is possible since $|\psi| \neq 0$. We have

$$v^i = \sum \phi_\rho^i(\tau)u^\rho.$$

As a special case, let us take the vector at ϵ with one in the ith component and zero elsewhere. From this vector we obtain the left-parallel vector

$$\mathbf{F}_i = (\phi_i^1, \cdots, \phi_i^r)' \qquad (\text{' denotes transpose}).$$

Thus: *each column of ϕ represents an infinitesimal transformation of the left-hand translation group, and the r columns form a basis of all infinitesimal transformations of the group.*

Let $\mathfrak{g}$ and $\mathfrak{h}$ be two abstract groups, $\mathfrak{G}$ and $\mathfrak{H}$ faithful realizations as groups of transformations on two spaces $\mathscr{S}$ and $\mathscr{T}$. Suppose $\mathfrak{G}$ and $\mathfrak{H}$ are isomorphic. We can extend this one-one correspondence to the infinitesimal transformations. That is, to each infinitesimal transformation $\mathbf{U}$ of $\mathfrak{G}$ there corresponds an infinitesimal transformation $\mathbf{V}$ of $\mathfrak{H}$. This correspondence is linear: $c_1 \mathbf{U}_1 + c_2 \mathbf{U}_2 \longrightarrow c_1 \mathbf{V}_1 + c_2 \mathbf{V}_2$. Furthermore, the commutator of two transformations of $\mathfrak{G}$ corresponds to the commutator of the two corresponding transformations of $\mathfrak{H}$: $[U_1 \, U_2] \longrightarrow [V_1 \, V_2]$. If we have merely a homomorphism $\mathfrak{G} \longrightarrow \mathfrak{H}$, then the correspondences go in one direction.

In particular, let $\mathbf{u}$ and $\mathbf{v}$ be two vectors attached to the identity ϵ of a group $\mathfrak{g}$. We extend this pair to the whole manifold by considering the vector fields of vectors left-parallel to $\mathbf{u}$ and $\mathbf{v}$. Thus we obtain two infinitesimal transformations of the left-hand translation group. The left translation group is isomorphic to any faithful realization of $\mathfrak{g}$, say $\mathfrak{G}$.

Let $\mathbf{u}_1, \mathbf{u}_2, \cdots, \mathbf{u}_r$ be a basis of the infinitesimal group. To each element $\mathbf{u}$ of this infinitesimal group corresponds an infinitesimal transformation $\mathbf{U}$ of the realization. To $\mathbf{u}_1, \mathbf{u}_2, \cdots, \mathbf{u}_r$ correspond $\mathbf{U}_1, \mathbf{U}_2, \cdots, \mathbf{U}_r$. We recall that

$$[\mathbf{u}_i \, \mathbf{u}_k] = \sum_\rho c_{ik}^\rho \mathbf{u}_\rho,$$

hence

$$[\mathbf{U}_i \, \mathbf{U}_k] = \sum_\rho c_{ik}^\rho \mathbf{U}_\rho.$$

Since any two columns of the matrix ϕ, say $\mathbf{F}_i$ and $\mathbf{F}_j$, are infinitesimal transformations of the left-hand translation group, it follows that the commutator $[\mathbf{F}_i \, \mathbf{F}_j]$ is a linear combination of the columns:

$$[\mathbf{F}_i \, \mathbf{F}_j] = \sum_\sigma c_{ij}^\sigma \mathbf{F}_\sigma.$$

In components this relation is

$$\sum_\sigma \left(\frac{\partial \phi_i^\rho}{\partial s^\sigma} \phi_k^\sigma - \frac{\partial \phi_k^\rho}{\partial s^\sigma} \phi_i^\sigma \right) = \sum_\sigma c_{ik}^\sigma \phi_\sigma^\rho.$$

30. First Fundamental Theorem of Lie Groups

We consider a vector ds_0 at the identity ϵ and a parallel vector ds at a point σ. Then as we saw, these vectors satisfy the relation

$$ds_0{}^i = \sum_{\rho=1}^r \psi_\rho^i(s) \, ds^\rho.$$

We also obtained infinitesimal transformations by considering "flows"

$$p = f[p_0, \sigma(t)],$$

or in terms of coordinates,

$$x^i = f^i(x_0, s), \qquad i = 1, 2, \cdots, n.$$

These flows give rise to the infinitesimal transformations of left-parallelism of vectors on $\mathfrak{g}$. We observed that two left-parallel elements give the same infinitesimal transformation. This leads to the fundamental equation

$$\sum_{\rho=1}^{r} \frac{\partial x^i}{\partial s^\rho}\, ds^\rho = \sum_{\rho=1}^{r} u_\rho^i(x)\, ds_0^\rho,$$

or

$$\sum_\rho \frac{\partial x^i}{\partial s^\rho}\, ds^\rho = \sum_{\rho,\sigma} u_\rho^i(x)\psi_\sigma^\rho\, ds^\sigma.$$

Equating coefficients of the differentials ds, we obtain

$$\frac{\partial x^i}{\partial s^k} = \sum_{\rho=1}^{r} u_\rho^i(x)\psi_k^\rho(s),$$

where $i = 1, 2, \cdots, n;\quad k = 1, 2, \cdots, r$. These equations express the First Fundamental Theorem of Sophus Lie. We repeat: they assert that two left-parallel vectors of $\mathfrak{g}$ yield the same infinitesimal transformation of the realized transformation group.

31. Mayer-Lie Systems

This system is a special case of a **Mayer-Lie system**, which in general has the form

$$\frac{\partial x^i}{\partial s^k} = F_k^i(x, s), \qquad i = 1, 2, \cdots, n;\quad k = 1, 2, \cdots, r.$$

Such a system for the n unknown functions $x^1, x^2, \cdots, x^n$ with r independent variables $s^1, \cdots, s^n$ will have solutions for an arbitrary choice of initial values x_0 at a given $s = s_0$ only if the $F_k^i(x, s)$ satisfy certain additional requirements. For example, for $n = 1$ and the F_k functions of s alone, we have

$$\frac{\partial x}{\partial s^k} = F_k(s), \qquad k = 1, 2, \cdots, r.$$

A necessary condition that such a system possess a solution is

$$\frac{\partial}{\partial s^j} F_k = \frac{\partial}{\partial s^k} F_j.$$

We shall now investigate conditions under which such Mayer-Lie systems possess solutions with arbitrarily prescribed initial values. Use will be made of this result when we discuss the converse of the so-called Second Fundamental Theorem of Lie.

To investigate the problem, we shall give it the following interpretation: We operate on the cartesian product $\mathscr{S} \times \mathfrak{g}$ of two manifolds $\mathscr{S}$ and $\mathfrak{g}$ of dimensions n and r, respectively, both smooth of at least the second order. To each line element on $\mathfrak{g}$, there corresponds linearly an infinitesimal transformation on $\mathscr{S}$ in the space of line elements attached to the same point on $\mathfrak{g}$. After introduction of coordinates $x^1, \cdots, x^n$ on $\mathscr{S}$ and $s^1, \cdots, s^r$ on $\mathfrak{g}$, the infinitesimal transformation corresponding to a line element ds^i will have components of the form

$$\sum_{\rho=1}^{r} F_\rho^i(x, s)\, ds^\rho, \qquad i = 1, 2, \cdots, n.$$

The $F_\rho^i(x, s)$ are assumed to be continuously differentiable on the product space $\mathscr{S} \times \mathfrak{g}$.

Let $\sigma(t)$, with components $s^i = s^i(t)$, be a curve in $\mathfrak{g}$. Then to each tangent vector of σ corresponds the infinitesimal transformation just described:

$$(\dagger) \qquad \frac{dx^i}{dt} = \sum_\rho F_\rho^i(x, s)\, \frac{ds^\rho}{dt}.$$

We compose such transformations by solving the system, getting a transformation which depends on the path σ. Evidently the resulting transformation is independent of the way in which the parameter t is chosen.

As an example we take $\mathscr{S}$ to be one-dimensional and $F = F(s)$. Then we have

$$\frac{dx}{dt} = \sum_\rho F_\rho(s)\, \frac{ds^\rho}{dt}.$$

By integration,

$$x = x_0 + \int_\alpha^\beta \sum_k F_k(s)\, ds^k,$$

and we see that in general the amount by which x is translated depends on the path. However if the F_k are the components of a gradient field, then the value of the integral is independent of the path.

We now investigate in general the circumstances under which the process described above is independent of the path. The complete integrability of $(\dagger)$ is obviously equivalent to this requirement. A necessary condition for independence is the independence of path in the two-dimensional subspaces in which

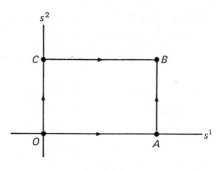

FIGURE 3

only two of the variables s^i vary. Let s^1, s^2 be the coordinates of such a subspace and consider the two paths OAB and OCB as shown in Figure 3.

According to the path OAB we obtain (by the above composition process) transformations satisfying the equations

$$\frac{\partial x^i}{\partial s^2} = F_2^i(x, s),$$

while on the path OCB,

$$\frac{\partial \bar{x}^i}{\partial s^1} = F_1^i(\bar{x}, s).$$

If the result is independent of the path, i.e., $x^i = \bar{x}^i$, then taking second derivatives yields

$$\frac{\partial^2 x^i}{\partial s^1 \, \partial s^2} = \sum_\rho \frac{\partial F_2^i}{\partial x^\rho} \frac{\partial x^\rho}{\partial s^1} + \frac{\partial F_2^i}{\partial s^1},$$

$$\frac{\partial^2 x^i}{\partial s^2 \, \partial s^1} = \sum_\rho \frac{\partial F_1^i}{\partial x^\rho} \frac{\partial x^\rho}{\partial s^2} + \frac{\partial F_1^i}{\partial s^2}.$$

Since

$$\frac{\partial x^\rho}{\partial s^1} = F_1^\rho, \qquad \frac{\partial x^\rho}{\partial s^2} = F_2^\rho,$$

we have

$$\sum_\rho \left(\frac{\partial F_2^i}{\partial x^\rho} F_1^\rho - \frac{\partial F_1^i}{\partial x^\rho} F_2^\rho \right) + \frac{\partial F_2^i}{\partial s^1} - \frac{\partial F_1^i}{\partial s^2} = 0.$$

We define

$$R_{kl}^i = \sum_\rho \left(\frac{\partial F_k^i}{\partial x^\rho} F_l^\rho - \frac{\partial F_l^i}{\partial x^\rho} F_k^\rho \right) + \left(\frac{\partial F_k^i}{\partial s^l} - \frac{\partial F_l^i}{\partial s^k} \right),$$

which may be called a **curvature quantity**. Thus a necessary condition for independence of path is $R_{kl}^i = 0$ identically in x and s for each i, k, l. In general we note that the R_{kl}^i, for each k and l form the components of an infinitesimal transformation.

More generally we can consider surfaces $s^i = s^i(p, q)_0$ and transport the quantities by two different paths between the same end points on such a surface. By considering the difference of the second derivatives along such paths, we find an expression of the form

$$\sum_{k, l} R_{kl}^i \frac{\partial s^k}{\partial p} \frac{\partial s^l}{\partial q}.$$

This expression is a generalization of the concept of the commutator of two infinitesimal transformations.

We shall prove in Lecture XII that the integrability conditions obtained are, at least in the small, not only necessary but also sufficient. Before we do this, we specialize to the case where the $F_k^i(x, s)$ have the form appearing in the equations expressing the First Fundamental Theorem of Lie:

$$F_k^i(x, s) = \sum_\rho u_\rho^i(x)\, \psi_k^\rho(s).$$

We assume that the infinitesimal transformations $\mathbf{U}_\rho$ on $\mathscr{S}$, with components $u_\rho^i(x)$, are linearly independent, and that the determinant $|\psi_k^\rho(s)|$ is different from zero everywhere on g. We compute the R_{kl}^i for these values of F_k^i:

$$\frac{\partial F_k^i}{\partial x^\rho} = \sum_\sigma \frac{\partial u_\sigma^i}{\partial x^\rho}\, \psi_k^\sigma, \qquad \frac{\partial F_k^i}{\partial s^l} = \sum_\sigma u_\sigma^i \frac{\partial \psi_k^\sigma}{\partial s^l}.$$

Hence,

$$R_{kl}^i = \sum_{\sigma, \rho, \tau} \left(\frac{\partial u_\sigma^i}{\partial x^\rho} \psi_k^\sigma u_\tau^\rho \psi_l^\tau - \frac{\partial u_\tau^i}{\partial x^\rho} \psi_l^\tau u_\sigma^\rho \psi_k^\sigma \right) + \sum_\sigma u_\sigma^i \left(\frac{\partial \psi_k^\sigma}{\partial s^l} - \frac{\partial \psi_l^\sigma}{\partial s^k} \right).$$

We denote by $[\mathbf{U}_\sigma\ \mathbf{U}_\tau]^i$, the ith component of the commutator $[\mathbf{U}_\sigma\ \mathbf{U}_\tau]$ of $\mathbf{U}_\sigma$ and $\mathbf{U}_\tau$. Then the above expression becomes

$$R_{kl}^i = \sum_{\sigma, \tau} \psi_k^\sigma \psi_l^\tau [\mathbf{U}_\sigma\ \mathbf{U}_\tau]^i + \sum_\sigma u_\sigma^i \left(\frac{\partial \psi_k^\sigma}{\partial s^l} - \frac{\partial \psi_l^\sigma}{\partial s^k} \right).$$

But we have

$$[\mathbf{U}_\sigma\ \mathbf{U}_\tau] = \sum_\rho c_{\sigma\tau}^\rho \mathbf{U}_\rho$$

with constant $c_{\sigma\tau}^\rho$, hence the condition for independence of path, the necessary condition for existence of a solution, is

$$\sum_{\rho, \sigma, \tau} c_{\sigma\tau}^\rho \psi_k^\sigma \psi_l^\tau \mathbf{U}_\rho = \sum_\sigma \left(\frac{\partial \psi_k^\sigma}{\partial s^l} - \frac{\partial \psi_l^\sigma}{\partial s^k} \right) \mathbf{U}_\sigma.$$

Since the $\mathbf{U}_\rho$ are assumed linearly independent, we can equate coefficients:

$$\frac{\partial \psi_k^\rho}{\partial s^l} - \frac{\partial \psi_l^\rho}{\partial s^k} = \sum_{\sigma, \tau} c_{\sigma\tau}^\rho \psi_k^\sigma \psi_l^\tau.$$

These equations, called **equations of Maurer-Cartan,** can easily be re-written as equations for the inverse matrix, $\|\phi_j^i\| = \|\psi_j^i\|^{-1}$, and yield the equations for the infinitesimal transformations represented by the columns of the ϕ matrix:

$$[\mathbf{F}_\sigma \ \mathbf{F}_\tau] = \sum_\rho c_{\sigma\tau}^\rho \mathbf{F}_\rho.$$

The interpretation of these equations as integrability conditions will be used in the proof of the converse of the Second Fundamental Theorem of Lie.

LECTURE XII

32. The Sufficiency Proof

We continue our study of the system

$$\frac{\partial x^i}{\partial s^k} = F_k^i(x, s),$$

where $x = (x^1, \cdots, x^n)$ is a point of a space $\mathscr{S}$ and $s = (s^1, \cdots, s^r)$ is a point of $\mathfrak{g}$. We noted that each contravariant vector u^k at a point s on $\mathfrak{g}$ determines an infinitesimal transformation $\sum \mathbf{F}_k u^k$ on $\mathscr{S}$. We may think of (s, u) as a line element at s on $\mathfrak{g}$, and thus we have a correspondence whereby to each line element on $\mathfrak{g}$ there corresponds an infinitesimal transformation on $\mathscr{S}$.

In particular, if $\sigma(t)$ is a differentiable curve on $\mathfrak{g}$ from α to β, then $d\sigma/dt$ (the velocity vector) determines such a line element at each point of $\sigma(t)$. We form

$$u^i(x, t) = \sum_k F_k^i(x, s(t)) \frac{ds^k}{dt},$$

an infinitesimal transformation on $\mathscr{S}$ for each fixed value of t. These transformations may be composed, i.e., integrated, yielding a finite transformation on $\mathscr{S}$. This is the process of solving the first-order system

$$\frac{dx^i}{dt} = u^i(x, t), \qquad x^i(0) = x^i_0,$$

which can always be done in the small by elementary results in differential equation theory. If $T[\sigma(t)]$ denotes the resulting transformation, then the corresponding infinitesimal transformation is precisely $u^i(x, t)$. In the case in which $\mathfrak{g}$ is a group with faithful realization $\sigma \longrightarrow T(\sigma)$, this resulting finite transformation is $T(\beta)\,T(\alpha)^{-1}$, which is independent of the path $\sigma(t)$.

In the general case of a Mayer-Lie system, we set

$$R^i_{kl} = \sum \left(\frac{\partial F^i_k}{\partial x^\rho} F^\rho_l - \frac{\partial F^i_\ell}{\partial x^\rho} F^\rho_k \right) + \left(\frac{\partial F^i_k}{\partial s^l} - \frac{\partial F^i_\ell}{\partial s^k} \right).$$

We proved in Lecture XI that if the process of composing the infinitesimal transformations corresponding to a curve in $\mathfrak{g}$ is independent of the path and only depends on the end points, then we have identically

$$R^i_{kl} = 0.$$

We shall prove the converse. We necessarily can obtain only a local result, since periods of integration may arise because of non-contractible closed curves in the large.

THEOREM 32.1. *If $R^i_{kl} = 0$ in a neighborhood of the origin, then the composition of infinitesimal transformations is independent of the path.*

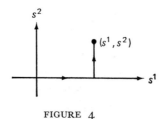

FIGURE 4

Proof. First we consider the case $r = 2$. We integrate over the path indicated in Figure 4 to obtain $x^i = x^i(x_0, s^1, s^2)$. It is clear that these functions satisfy

$$\frac{\partial x^i}{\partial s^2} = F^i_2(x, s)$$

in a neighborhood of 0. We now consider the equations

$$\frac{\partial x^i}{\partial s^1} - F^i_1(x, s) = \Delta^i(x_0, s^1, s^2),$$

$$\frac{\partial x^1}{\partial s^2} - F^i_2(x, s) = 0.$$

We shall prove that the so-defined functions Δ^i vanish identically. First we note that

($\dagger$) $$\Delta^i(x_0, s^1, 0) = 0,$$

since for $s^2 = 0$, the function x^i is defined by an integration along the s^1-axis.

88   *Lecture XII*

We have

$$\frac{\partial \Delta^i}{\partial s^2} = \frac{\partial^2 x^i}{\partial s^1 \, \partial s^2} - \sum_\rho \frac{\partial F_1^i}{\partial x^\rho} F_2^\rho - \frac{\partial F_1^i}{\partial s^2}.$$

But

$$\frac{\partial^2 x^i}{\partial s^1 \, \partial s^2} = \frac{\partial}{\partial s^1} (F_2^i(x, s)) = \sum_\rho \frac{\partial F_2^i}{\partial x^\rho} \frac{\partial x^\rho}{\partial s^1} + \frac{\partial F_2^i}{\partial s^1}$$

$$= \sum_\rho \frac{\partial F_2^i}{\partial x^\rho} (\Delta^\rho + F_1^\rho) + \frac{\partial F_2^i}{\partial s^1}$$

$$= \sum_\rho \frac{\partial F_2^i}{\partial x^\rho} \Delta^\rho + \sum_\rho \frac{\partial F_2^i}{\partial x^\rho} F_1^\rho + \frac{\partial F_2^i}{\partial s^1},$$

therefore

$$\frac{\partial \Delta^i}{\partial s^2} = \sum_\rho \left(\frac{\partial F_2^i}{\partial x^\rho} F_1^\rho - \frac{\partial F_1^i}{\partial x^\rho} F_2^\rho \right) + \left(\frac{\partial F_2^i}{\partial s^1} - \frac{\partial F_1^i}{\partial s^2} \right)$$

$$+ \sum_\rho \frac{\partial F_2^i}{\partial x^\rho} \Delta^\rho = R_{21}^i + \sum_\rho \frac{\partial F_2^i}{\partial x^\rho} \Delta^\rho.$$

By our hypothesis we finally have

$$\frac{\partial \Delta^i}{\partial s^2} = \sum_\rho \frac{\partial F_2^i}{\partial x^\rho} \Delta^\rho.$$

We hold s^1 constant; then this may be considered as a system of homogeneous first order equations for the Δ^i. The initial conditions (†) combined with the uniqueness theorem for a first-order system of ordinary differential equations imply that $\Delta^i = 0$.

It follows that there are functions $x^i = x^i(x_0, s^1, s^2)$ such that

$$\frac{\partial x^i}{\partial s^1} = F_1^i(x, s), \quad \frac{\partial x^i}{\partial s^2} = F_2^i(x, s),$$

and

$$x^i(x_0, 0, 0) = x_0^i.$$

It is evident from this that integration along any curve $s(t)$ from 0 to (s^1, s^2) would yield the same result.

For the higher dimensional case, we integrate from 0 to $(s^1, \cdots, s^r)$ along a polygonal path of segments always parallel to a coordinate axis. There are $r!$ such paths corresponding to the possible permutations of the axes; however each permutation is composed of a sequence of transpositions of adjacent positions, and we know from the case $r = 2$ that interchange of adjacent variables does not change the result of the integration. Thus each of the $r!$ paths yields the same answer; since we may always select such a path in which the final segment is parallel to an arbitrary one of the coordinate axes, we obtain functions $x^i = x^i(x_0, s)$ such that

$$\frac{\partial x^i}{\partial s^k} = F_k^i(x, s), \quad x^i(x_0, 0) = x_0^i.$$

Evidently this insures that the integration will be independent of path.

33. First Fundamental Theorem; Converse

We return to the case of a group $\mathfrak{g}$ having a faithful realization on $\mathscr{S}$. We had

(*) $$F_k^i = \sum u_\rho^i(x)\, \psi_k^\rho(s),$$

where $|\psi_k^\rho| \neq 0$ and $\mathbf{U}_\rho = (u_\rho^1, \cdots, u_\rho^n)'$ are the infinitesimal transformations of $\mathfrak{g}$ on $\mathscr{S}$. The integrability conditions $R_{kl}^i = 0$ reduced to the equations

$$[\mathbf{U}_k\ \mathbf{U}_l] = \sum c_{kl}^\rho \mathbf{U}_\rho$$

and to the Maurer-Cartan equations

$$\frac{\partial \psi_k^i}{\partial s^l} - \frac{\partial \psi_l^i}{\partial s^k} = \sum c_{\rho\sigma}^i \psi_k^\rho \psi_l^\sigma.$$

The equations (*) constitute the First Fundamental Theorem; we shall state a converse.

THEOREM 33.1. *Let $\mathfrak{g}$ be an r-dimensional connected manifold, $\mathscr{S}$ an n-dimensional manifold, and $\sigma \longrightarrow T(\sigma)$ a smooth mapping which takes each element σ of $\mathfrak{g}$ to a transformation on $\mathscr{S}$. We make the following assumptions:*

1. *If $\sigma_1 \neq \sigma_2$, then $T(\sigma_1) \neq T(\sigma_2)$.*
2. *If $x^1, \cdots, x^n$ are coordinates at p, if $s^1, \cdots, s^r$ are coordinates at σ, and if $p = T(\sigma)p_0$, then*

$$\frac{\partial x^i}{\partial s^k} = \sum_\rho u_\rho^i(x)\psi_k^\rho(s),$$

where $|\psi_k^\rho| \neq 0$, and $\mathbf{U}_1, \cdots, \mathbf{U}_r$ are independent infinitesimal transformations on $\mathscr{S}$, $\mathbf{U}_\rho = (u_\rho^1, \cdots, u_\rho^n)'$.

3. *There is an ϵ in $\mathfrak{g}$ such that $T(\epsilon) = I$.*

4. *If $\sigma(t)$ is a differentiable curve in $\mathfrak{g}$ from α to β and α_1 is any point of $\mathfrak{g}$, then there exists a curve $\sigma_1(t)$ starting at α_1 for which parallelism of tangents is preserved,*

i.e., such that

$$\sum \psi_k^\rho(s) \frac{ds^k}{dt} = \sum \psi_k^\rho(s_1) \frac{ds_1^k}{dt},$$

where $\sigma(t)$ and $\sigma_1(t)$ have respective parameterizations $s^k(t)$ and $s_1^k(t)$. Then $\mathfrak{g}$ is a group, the functions ψ_k^ρ represent the parallelism on $\mathfrak{g}$, $\sigma \longrightarrow T(\sigma)$ is a faithful realization of $\mathfrak{g}$, and the infinitesimal transformations of $\mathfrak{g}$ on $\mathscr{S}$ have basis $\mathbf{U}_1, \cdots, \mathbf{U}_r$.

Discussion. It is clear why we have Assumptions 1 and 2. It is unnecessary to impose integrability conditions on the system in Assumption 2 because, presupposing that a solution exists, they follow automatically. Assumption 3 fixes an identity element in $\mathfrak{g}$. We note that if T_0 is a fixed transformation of $\mathscr{S}$ and we replace $T(\sigma)$ by $T^*(\sigma) = T(\sigma) T_0$, then the other hypotheses are still satisfied, so that they alone do not determine an identity in $\mathfrak{g}$. Assumption 4 has the purpose of making $\mathfrak{g}$ a full group. In fact any open connected neighborhood of ϵ in a group will satisfy all of the hypotheses except Assumption 4, so that this is essential if we are to conclude that $\mathfrak{g}$ is a full group. In the situation of Assumption 4 in which α and the path $\sigma(t)$ from α to β are given, we may multiply by $\|\psi_k^\rho(s_1)\|^{-1}$ to express Assumption 4 in the form

$$\frac{ds_1^k}{dt} = h^k(t, s_1).$$

For definiteness, let us suppose the time interval is $[0, 1]$ so that $\sigma(0) = \alpha$, $\sigma(1) = \beta$. We seek $\sigma_1(t)$ with the prescribed initial condition $\sigma_1(0) = \alpha_1$ and satisfying this first-order system. General theory tells us that a solution exists on part of the interval, but not necessarily on the whole interval. Assumption 4 guarantees that the solution exists on the whole time interval.

Proof. Given $\alpha, \beta \in \mathfrak{g}$, we wish to find $\gamma \in \mathfrak{g}$ such that $T(\beta) T(\alpha) = T(\gamma)$. Since γ, if it exists, is unique by Assumption 1, we may define the product $\beta\alpha = \gamma$. See Figure 5. To construct γ, we connect ϵ to α by a smooth curve. To each tangent vector of this path corresponds an infinitesimal transformation on $\mathscr{S}$, and $T(\alpha)$ is composed of these. We also take a smooth path from ϵ to β.

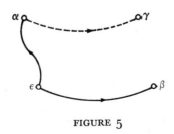

FIGURE 5

By Assumption 4, there is a path from α to an endpoint we shall call γ, this path being parallel to the one from ϵ to β. We see that $T(\gamma)$ is obtained by composing the line elements from ϵ to α and then from α to γ. However, the composition from α to γ is exactly the same as that from ϵ to β according to Assumption 2; hence

$$T(\gamma) = [T(\gamma)\,T(\alpha)^{-1}][T(\alpha)] = T(\beta)\,T(\alpha).$$

Thus we have defined a multiplication operation on $\mathfrak{g}$. It remains to get the inverse. If $\alpha \in \mathfrak{g}$, we connect α to ϵ by a smooth curve (Figure 6) and then take a parallel curve from ϵ to α'. Since the infinitesimal transformations along the curve from α to ϵ are the negatives of those along the same curve, but in the opposite direction, it follows that the composed transformation $T(\alpha')$ is precisely $T(\alpha)^{-1}$. The remaining conclusions are evident.

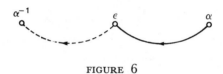

FIGURE 6

34. Second Fundamental Theorem; Converse

Suppose that $\mathscr{S}$ is an n-dimensional manifold and that $\mathbf{U}_1, \cdots, \mathbf{U}_r$ are linearly independent infinitesimal transformations on $\mathscr{S}$. We ask whether they are the infinitesimal transformations of some group. An evident necessary condition is

$$[\mathbf{U}_i\,\mathbf{U}_k] = \sum c_{ik}^\rho U_\rho, \qquad c_{ik}^\rho \text{ constants.}$$

Another necessary condition is obtained as follows. Let $a^1, \cdots, a^r$ be constants and consider the infinitesimal transformation $\sum a^i\,\mathbf{U}_i$. It defines a steady flow on $\mathscr{S}$; if we actually had a group, this flow would yield one-one transformations on $\mathscr{S}$. In other words, the system

$$\frac{dp}{dt} = \sum a^i\,\mathbf{U}_i$$

could be integrated along any path and would yield a transformation of $\mathscr{S}$. We shall see later that these necessary conditions are also sufficient. The problem is to find a space $\mathfrak{g}$ and functions $\psi_k^\rho(s)$ such that $F_k^i = \sum u_\rho^i(x)\psi_k^\rho(s)$ belongs to an integrable system, i.e.,

$$\frac{\partial \psi_k^i(s)}{\partial s^l} - \frac{\partial \psi_l^i(s)}{\partial s^k} = \sum c_{\rho\sigma}^i \psi_k^\rho \psi_l^\sigma.$$

NOTE. Lie developed the theory of contact in order to solve this system of equations, where the $c^i_{\rho\sigma}$ are arbitrary skew-symmetric constants that satisfy the quadratic relations which amount to Jacobi's identity.

We observe that these are invariant equations, hence we may specialize the coordinate system. To see how to do so effectively, we return to the case of a group and introduce **canonical coordinates** near ϵ. To each parallel infinitesimal transformation on $\mathfrak{g}$ we associate a path through ϵ by composing the field, or what is the same thing, solving

$$\frac{d\sigma}{dt} = \mathbf{a}\sigma,$$

where $\mathbf{a}$ is a fixed vector at ϵ, and is $\mathbf{a}\sigma$ is the translate of $\mathbf{a}$ to σ. Thus there is a curve through ϵ in each direction. We introduce coordinates so as to make these curves become straight lines.

Clearly a neighborhood of ϵ is covered by these curves. However even when $\mathfrak{g}$ is connected, the whole group $\mathfrak{g}$ is not necessarily covered, as we see from the following example:

Example. $\mathfrak{g}$ the real 2×2 unimodular group consisting of real matrices

$$A = \begin{pmatrix} a_{11} & a_{12} \\ a_{21} & a_{22} \end{pmatrix}, \qquad |A| = 1.$$

It is a connected three-parameter group. The infinitesimal transformations are the matrices

$$B = \begin{pmatrix} b_{11} & b_{12} \\ b_{21} & b_{22} \end{pmatrix}$$

of trace zero. The curve through $\epsilon = I$ corresponding to such a B is given by

$$e^{Bt}.$$

We wish to show that not every A is on one of these curves. Evidently it suffices to show that the equation $A = e^B$ is impossible for some A. We do this by finding the characteristic roots of e^B. Since B has trace 0, the characteristic equation of B is

$$x^2 + |B| = 0,$$

so that the roots of B are either $\lambda = \alpha \geq 0$, $\mu = -\alpha$, or $\lambda = i\alpha$, $\mu = -i\alpha$, $\alpha \geq 0$. The roots of e^B are $\rho = e^\lambda$ and $\sigma = e^\mu$. Thus $\rho\sigma = 1$, and in the first case $\rho > 0$, $\sigma > 0$, while in the second case $|\rho| = |\sigma| = 1$. It remains to write an element of the real unimodular group whose characteristic roots do not satisfy these conditions. Such a matrix is

$$A = \begin{pmatrix} -\frac{1}{2} & 0 \\ 0 & -2 \end{pmatrix}.$$

LECTURE XIII

35. Converse of Second Fundamental Theorem (continued)

Let U_1, U_2, $\cdots$, U_r be linear infinitesimal transformations on an n-dimensional manifold. Suppose that the commutators satisfy the relations

$$[U_k\, U_l] = \sum_{\rho=1}^{r} c^{\rho}_{kl}\, U_\rho.$$

We naturally assume that the constants c^{ρ}_{kl} are antisymmetric in the lower indices, and that the constants $c^{i}_{klm} = \sum_\rho c^{i}_{\rho k} c^{\rho}_{lm}$ satisfy the Jacobi identity. We inquire under what additional hypotheses are the U_ρ the infinitesimal transformations of a group.

From the converse of the First Fundamental Theorem the problem becomes one of solving the system

(1)
$$\frac{\partial \psi^i_k}{\partial s^l} - \frac{\partial \psi^i_l}{\partial s^k} = \sum_{\rho,\sigma} c^{i}_{\rho\sigma}\, \psi^\rho_k(s)\, \psi^\sigma_l(s),$$

where $\det (\psi^i_k) \neq 0$ on the entire manifold. For this purpose we introduce canonical coordinates as described in Lecture XII.

In order to solve the system we introduce parameters

$$s^i = a^i t,$$

and add to (1) the conditions

$$\sum \psi^i_\rho(s)\, s^\rho = s^i.$$

These mean that if the U_k are the infinitesimal transformations of a group, then the coordinates s^i are canonical coordinates in this group.

We first note that

$$\frac{\partial \psi_k^i}{\partial t} = \sum_\rho \frac{\partial \psi_k^i}{\partial s^\rho} a^\rho,$$

and define

$$X_k^i = t \psi_k^i.$$

Then

$$\frac{\partial X_k^i}{\partial t} = \sum_\rho \frac{\partial \psi_k^i}{\partial s^\rho} s^\rho + \psi_k^i.$$

Multiplying (1) by s^l and summing on the index l, we have

$$\sum_l \left(\frac{\partial \psi_k^i}{\partial s^l} s^l - \frac{\partial \psi_l^i}{\partial s^k} s^l \right) = \sum_{\rho,\sigma,l} c_{\rho\sigma}^i \, \psi_k^\rho(s) \, \psi_l^\sigma(s) \, s^l,$$

$$\frac{\partial X_k^i}{\partial t} - \psi_k^i - \sum_l \frac{\partial (\psi_l^i s^l)}{\partial s^k} + \psi_k^i = \sum_{\rho\sigma} c_{\rho\sigma}^i \psi_k^\rho s^\sigma.$$

Noting that

$$\sum_\ell \frac{\partial (\psi_l^i s^l)}{\partial s^k} = \frac{\partial s^i}{\partial s^k} = \delta_k^i$$

and setting

$$h_\rho^i = \sum_\sigma c_{\rho\sigma}^i a^\sigma,$$

we have

(2)
$$\frac{\partial X_k^i}{\partial t} = \delta_k^i + \sum_\rho h_\rho^i X_k^\rho.$$

For fixed $a^1, a^2, \cdots, a^r$ the system (2) is a system of ordinary linear differential equations for the unknown functions X_k^ρ as functions of t, subject to the initial conditions

$$X_k^i = 0 \qquad \text{for} \qquad t = 0.$$

Hence by the general theory of ordinary differential equations there is a solution analytic in the entire space. We must demonstrate that the resulting functions ψ_k^i are actually functions of $s^1, s^2, \cdots, s^r$. But if we replace t by ct, a by a/c and X by cX, the equations (2) remain unchanged. Hence the functions ψ_k^i are functions of $a^i t$, i.e., functions of s.

It remains to be shown that solutions of (2) yield functions ψ_k^i which satisfy (1). From the definition of X_k^i,

$$\frac{\partial \psi_k^i}{\partial a^l} = \frac{\partial \psi_k^i}{\partial s^l} t, \quad \frac{\partial X_k^i}{\partial a^l} = \frac{\partial \psi_k^i}{\partial s^l} t^2.$$

We define the functions V^i_{kl} by the relations

$$V^i_{kl} = \frac{\partial X^i_k}{\partial a^l} - \frac{\partial X^i_l}{\partial a^k} - \sum_{\rho, \sigma} c^i_{\rho\sigma} X^\rho_k X^\sigma_l.$$

The right side set equal to zero is nothing but equation (1) multiplied by t^2. Thus we must show that if X^i_k is a solution of (2), then the V^i_{kl} vanish.

We differentiate with respect to t:

$$\frac{\partial V^i_{kl}}{\partial t} = \frac{\partial}{\partial a^l}\left(\frac{\partial X^i_k}{\partial t}\right) - \frac{\partial}{\partial a^k}\left(\frac{\partial X^i_l}{\partial t}\right) - \sum_{\rho, \sigma} c^i_{\rho\sigma} \frac{\partial X^\rho_k}{\partial t} X^\sigma_l - \sum_{\rho, \sigma} c^i_{\rho\sigma} X^\rho_k \frac{\partial X^\sigma_l}{\partial t}$$

$$= \sum_\rho \left(\frac{\partial h^i_\rho}{\partial a^l} X^\rho_k - \frac{\partial h^i_\rho}{\partial a^k} X^\rho_l\right) - \sum_{\rho, \sigma}(c^i_{\rho\sigma}\delta^\rho_k X^\sigma_l + c^i_{\rho\sigma} X^\rho_k \delta^\sigma_l)$$

$$+ \sum_\rho \left(h^i_\rho \frac{\partial X^\rho_k}{\partial a^l} - h^i_\rho \frac{\partial X^\rho_l}{\partial a^k}\right) - \sum_{\rho, \sigma, \omega}(c^i_{\rho\sigma} h^\rho_\omega X^\omega_k X^\sigma_l + c^i_{\rho\sigma} X^\rho_k h^\sigma_\omega X^\omega_l).$$

Inserting

$$\frac{\partial h^i_\rho}{\partial a^l} = c^i_{\rho l}$$

into the first two terms above and taking account of the anti-symmetry of the $c^i_{\rho\sigma}$, we see that the first four terms cancel. Then by substituting $h^\rho_\omega = c^\rho_{\omega\lambda} a^\lambda$ and by using the cyclic relations (Section 24) on the c^i_{klm}, we simplify the remaining terms. We finally get

$$\frac{\partial V^i_{kl}}{\partial t} = \sum_\rho h^i_\rho V^\rho_{kl}.$$

But $V^i_{kl} = 0$ for $t = 0$, hence the first-order homogeneous system for the V^i_{kl} has only the identically zero solution.

36. Concept of Group Germ

We obtain a new concept, that of group germ, when we abstract the properties of any neighborhood of the identity in a Lie group.

DEFINITION. A **group germ** consists of a neighborhood $\mathcal{N}$ of the origin in $\mathcal{E}^n$ together with an operation of composition $\circ$ which maps certain elements of $\mathcal{N} \times \mathcal{N}$ into $\mathcal{N}$ according to the following rules:

1. If either $a \circ (b \circ c)$ or $(a \circ b) \circ c$ is defined, so is the other and they are equal.

2. If $a \circ b = c$, then there is a neighborhood $\mathcal{U}$ of a and a neighborhood $\mathcal{V}$ of b such that $x \circ y$ is defined for each $x \in \mathcal{U}, y \in \mathcal{V}$.

3. If we represent the origin of $\mathscr{E}^n$ by ϵ, then $x \circ \epsilon$ and $\epsilon \circ x$ are defined for each $x \in N$ and $x \circ \epsilon = \epsilon \circ x = x$.

4. There is a neighborhood of ϵ in $\mathcal{N}$ in which each element x has a unique inverse x^{-1}: $x \circ x^{-1} = x^{-1} \circ x = \epsilon$.

5. The functions $(x, y) \longrightarrow x \circ y$ and $x \longrightarrow x^{-1}$ are continuous of class C^3 where defined.

From an arbitrarily small neighborhood of the identity on a given Lie group, we can obtain all of the infinitesimal transformations (left parallel fields), hence the Lie algebra and the constants of structure. We easily see that the same applies on a group germ (also called a **local Lie group**), so we may define the Lie algebra of a group germ and its constants of structure.

We now pose this problem. Given n^3 constants c_{jk}^i, antisymmetric in j, k, that satisfy the relations

$$c_{klm}^i + c_{lmk}^i + c_{mkl}^i = 0,$$

where

$$c_{klm}^i = \sum_\rho c_{\rho k}^i c_{lm}^\rho,$$

does there exist a Lie group for which these are the constants of structure? Or stated another way, given a Lie algebra, does there exist a corresponding Lie group? The answer is yes, but it is beyond the time limitations of these lectures to prove this in the general case; we shall discuss it somewhat below. For the moment, however, let us consider the less stringent question: given constants c_{kl}^i as above, does there exist a group germ for which these are the constants of structure? This is the case, and the proof of the preceding section shows us exactly this. Briefly, if the c_{kl}^i are given, we obtain functions $\psi_k^i(x)$ on a neighborhood $\mathcal{N}$ of the origin in $\mathscr{E}^n$, with non-vanishing determinant, and satisfying

$$\frac{\partial \psi_k^i}{\partial x^l} - \frac{\partial \psi_l^i}{\partial x^k} = \sum_{\rho, \sigma} c_{\rho\sigma}^i \, \psi_k^\rho(x) \, \psi_l^\sigma(x).$$

If we define fields $\mathbf{U}_1, \cdots, \mathbf{U}_n$ by

$$\mathbf{U}_i = (\phi_i^1, \cdots, \phi_i^n)', \quad \sum_\sigma \psi_\sigma^i \phi_k^\sigma = \delta_k^i,$$

then

$$[\mathbf{U}_i \, \mathbf{U}_k] = \sum c_{ik}^\rho \mathbf{U}_\rho,$$

and we simply apply a local version of the converse to the First Fundamental Theorem.

The more serious problem of constructing a whole Lie group from a given Lie algebra is, as noted before, a more difficult one. It turns out to be quite easy in several special cases; we shall consider first the special case in which the center of the given Lie algebra vanishes.

LECTURE XIV

37. Converse of the Third Fundamental Theorem

We were considering the case in which the center of the Lie algebra Γ is 0. In general, we introduce the adjoint group defined by $\sigma \longrightarrow \alpha\sigma\alpha^{-1}$ for each fixed α in $\mathfrak{g}$ and pass to the linear adjoint group by considering the transformations induced on the space of vectors at ϵ. This is a representation of $\mathfrak{g}$.

Now suppose we are given an abstract Lie algebra Γ. This consists of an r-dimensional linear space with a distributive product $[\mathbf{u}\ \mathbf{v}]$ such that

$$(1) \qquad\qquad [\mathbf{u}\ \mathbf{v}] = -[\mathbf{v}\ \mathbf{u}],$$

$$(2) \qquad\qquad [[\mathbf{u}\ \mathbf{v}]\ \mathbf{w}] + [[\mathbf{v}\ \mathbf{w}]\ \mathbf{u}] + [[\mathbf{w}\ \mathbf{u}]\ \mathbf{v}] = 0.$$

Let $\mathbf{a}_1, \cdots, \mathbf{a}_r$ be a linear basis of Γ. Then

$$[\mathbf{a}_i \mathbf{a}_j] = \sum_{\rho} c_{ij}^{\rho} \mathbf{a}_{\rho},$$

where the c's are anti-symmetric in i and j because of (1), and satisfy the usual quadratic relations because of the Jacobi identity (2). The problem is to show that the c's are derived from a group.

We have already proved that if the c's are derived from a group, then the adjoint representation is given as follows:

$$(3) \qquad\qquad A(\mathbf{a}) \colon \mathbf{u} \longrightarrow [\mathbf{u}\ \mathbf{a}].$$

It follows that

$$A(\mathbf{a})A(\mathbf{b}) - A(\mathbf{b})A(\mathbf{a}) = A([\mathbf{a}\ \mathbf{b}]).$$

This relation is equivalent to the Jacobi identity, and it asserts that the commutator of $A(\mathbf{a})$ and $A(\mathbf{b})$ is $A([\mathbf{a}\ \mathbf{b}])$. It follows that if we set $A_i = A(\mathbf{a}_i)$, then

$$[A_i\ A_j] = \sum_\rho c_{ij}^\rho A_\rho.$$

In other words, we obtain r infinitesimal transformations of the linear group in the r-dimensional space.

We can now piece together a converse to the Third Fundamental Theorem. Suppose Γ is an r-dimensional Lie algebra with a basis $\mathbf{a}_1, \cdots, \mathbf{a}_r$. We suppose the center of Γ is 0. The correspondence $\mathbf{a} \longrightarrow A(\mathbf{a})$, where $A(\mathbf{a})$ is defined by (3), takes each element $\mathbf{a}$ of Γ onto an infinitesimal transformation of the linear group of r-variables. If $\mathbf{a}_i \longrightarrow A_i$, then we assert that $A_1, \cdots, A_r$ *are linearly independent.* For suppose $\sum c^i A_i = 0$. Then

$$\left[\mathbf{u} \sum c^i \mathbf{a}_i\right] = 0$$

for all $\mathbf{u}$. Thus $\sum c^i \mathbf{a}_i$ is in the center of Γ, and it must vanish. This implies $c^i = 0$ since $\mathbf{a}_1, \cdots, \mathbf{a}_r$ is a basis of Γ.

We previously proved: Given r linearly independent infinitesimal transformations $A_1, \cdots, A_r$ on a space $\mathscr{S}$, closed under $[\ ]$, then there exists a transformation group generated by the given infinitesimal transformations. We used one additional assumption: any linear combination $\sum b^i A_i$ can be integrated. This is certainly satisfied in the present case because each linear combination of the A_i is an infinitesimal of a group, namely the full linear group. Thus we do obtain a group $\mathfrak{g}$ from the given Lie algebra Γ of center zero.

Time does not allow our completing the converse of the Third Fundamental Theorem; we sketch the technique. Let Γ be a Lie algebra and let Γ' be the linear hull of all brackets $[\mathbf{u}\ \mathbf{v}]$ where $\mathbf{u}, \mathbf{v} \in \Gamma$. We call Γ' the **derived algebra** of Γ. Thus we have $\Gamma'' = (\Gamma')'$, Γ''', etc., and evidently

$$\Gamma \supseteq \Gamma' \supseteq \Gamma'' \supseteq \cdots.$$

The Lie algebra Γ is called **integrable** or **solvable** if this sequence eventually comes to 0. We note that if $\Gamma^{(r)} = \Gamma^{(r+1)}$ for some r, then

$$\Gamma^{(r+1)} = \Gamma^{(r+2)} = \Gamma^{r(+3)} = \cdots,$$

so that if Γ is solvable we must have

$$\Gamma \supset \Gamma' \supset \Gamma'' \supset \cdots \supset \Gamma^{(r)} \supset \Gamma^{(r+1)} = 0.$$

It turns out that if Γ is solvable, the Converse of the Third Fundamental Theorem is relatively easy to prove. The general case is then obtained by a deep structure theorem which asserts that any Lie algebra Γ may be decomposed into the sum of a solvable aglebra and an algebra with zero center.

38. The Helmholtz-Lie Problem

We shall conclude these lectures with a discussion of the following problem: to characterize in a group-theoretic manner the three groups of elementary geometric spaces:

1. Euclidean space,
2. Bolyai-Lobachevski hyperbolic geometry,
3. Spherical space.

Let $\mathscr{S}$ be one of these spaces, $\mathfrak{G}$ the corresponding group. If ω is a one-one correspondence on $\mathscr{S}$ onto a space $\mathscr{T}$, then $\mathfrak{G}$ gives rise to a similar group $\mathfrak{H}$ on $\mathscr{T}$. There is no essential distinction between $\mathfrak{G}$ and $\mathfrak{H}$, the only difference is that the elements of the underlying space $\mathscr{S}$ are relabeled. Consequently we shall seek a system of axioms that characterizes the groups of the spaces 1–3 up to similarity.

Let us first consider the two-dimensional case. We begin with the Euclidean plane with its group of proper motions. We consider the set of objects consisting of a point in the plane and a directed line element through the point. Each motion sends such a line element to another line element. Hence the group of proper motions of the Euclidean plane induces a group of transformations on the space of line elements. This extended group is transitive, because by a translation we may carry a point p to a point q and by a rotation around q we can carry any directed line element at q to any other. What is more, there is only one proper motion sending one of a given pair of directed line elements to another. Hence the extended group is **simply transitive** in the space of directed line elements.

Evidently the same remarks apply to the group of proper motions of the unit sphere $\mathscr{S}^2$ in the Euclidean space $\mathscr{E}^3$.

The Bolyai-Lobachevskk geometry of two dimensions can be represented by the open unit circle $|z| < 1$ in the complex plane with the group of Möbius transformations

$$w = \epsilon \frac{z - \alpha}{1 - \bar{\alpha} z}, \qquad |\alpha| < 1, \quad |\epsilon| = 1.$$

This is a three-parameter group and again its action is simply transitive on the space of directed line elements.

The Helmholtz-Lie problem has been formulated for two dimensions by H. Weyl as follows: Find all groups $\mathfrak{G}$ on a two-dimensional manifold that act in a simply transitive way on the space of directed line elements.

In three dimensions, we introduce the notion of a flag (so-named by

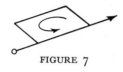

FIGURE 7

Kowalewski). A **flag** consists of a point, a directed line on the point, and an oriented plane element on the line (Figure 7).

Suppose we consider the group of proper motions acting in the Euclidean space $\mathscr{E}^3$. This is a six-dimensional group and has the property of acting in a simply transitive way on the flags: If two flags are given, there exists a unique proper motion of $\mathscr{E}^3$ sending one onto the other.

The same transitivity is true of the group of proper motions of the three-sphere $\mathscr{S}^3 \subset \mathscr{E}^4$, which is defined as usual by $x_1^2 + \cdots + x_4^2 = 1$. Also the Bolyai-Lobachevski three-dimensional hyperbolic geometry has the property. As a model for this last space, we may take the open unit ball $x_1^2 + x_2^2 + x_3^2 < 1$ in $\mathscr{E}^3$ for the space and the set of Möbius transformations that send this region onto itself for the group.

We now define the notion of **flag** in n-dimensions. This consists of a point, a directed line on the point, an oriented plane element on the line, an oriented three-dimensional element on the plane, $\cdots$, and finally an oriented $(n-1)$-element on the $(n-2)$-element. (We could also add an oriented n-element, but this would only have the effect of including the improper motions in the picture.)

Our problem is this: Let $\mathscr{S}$ be an n-dimensional manifold and $\mathfrak{G}$ a group of transformations on $\mathscr{S}$, where everything is assumed sufficiently differentiable. Assume that $\mathscr{S}$ is connected and that $\mathfrak{G}$ acts in a simply transitive way on the space of all flags on $\mathscr{S}$. We wish to characterize such a group $\mathfrak{G}$. We remark at the outset that the hypotheses imply that $\mathfrak{G}$ is a space of dimension $n(n+1)/2$ since this number is the dimension of the space of flags on $\mathscr{S}$. We shall show that with one exception, $\mathfrak{G}$ must be one of the groups of elementary geometry, already enumerated. We shall also show that these various groups are not themselves mutually similar.

We note that the dimension of $\mathfrak{G}$ does not characterize $\mathfrak{G}$ in itself. In fact the group

$$x' = ax + by$$
$$y' = cy \qquad a > 0, \quad c > 0,$$

of **Galilean transformations** of the plane has dimension three but is not similar to one of the groups of elementary geometry since it does not act transitively on the directed line elements. (A flag parallel to the x-axis remains so under any transformation of the group.)

We fix a point O of $\mathscr{S}$ and consider the subgroup $\mathfrak{G}_0$ of $\mathfrak{G}$ consisting of all T in $\mathfrak{G}$ such that $T(O) = O$. It is clear that $\mathfrak{G}_0$ induces a linear group $\mathfrak{L}_0$ on the tangent space to $\mathscr{S}$ at O, so that we have a representation $\mathfrak{G}_0 \longrightarrow \mathfrak{L}_0$.

The set of flags at O forms a compact space of $n(n-1)/2$ parameters; given any two flags at O, there is a unique transformation in $\mathfrak{L}_0$ sending one to the other. This evidently means that $\mathfrak{L}_0$ is topologically equivalent to the rotation group in $\mathscr{E}^n$, hence $\mathfrak{L}_0$ is compact. We note, incidentally that $\mathfrak{G}_0 \longrightarrow \mathfrak{L}_0$ is a faithful representation, so we also have proved that $\mathfrak{G}_0$ *is compact.* At any rate, we know that *there exists an invariant positive definite quadratic form* $f = \sum a_{\rho\sigma} u^\rho u^\sigma$, where u^ρ denotes the typical vector in the tangent space at O. We mean to say that f is invariant under $\mathfrak{L}_0$. (This was proved in Lecture VII.) We shall prove that *the form f is unique up to a positive constant factor.* In fact, suppose $\sum b_{\rho\sigma} u^\rho u^\sigma$ is another invariant form, not proportional to f. We select an eigenvalue λ of $\|a_{\rho\sigma}\|$ with respect $\|b_{\rho\sigma}\|$ and have

$$\sum_\rho a_{i\rho} u^\rho = \sum_\rho \lambda b_{i\rho} u^\rho,$$

where $\mathbf{u}$ is some non-zero vector. The totality of such eigenvectors $\mathbf{u}$ form a proper linear subspace of the tangent space at O. This subspace is invariant under $\mathfrak{L}_0$ since the quadratic forms are; hence $\mathfrak{L}_0$ is reducible. But this is impossible because $\mathfrak{L}_0$ acts transitively on the flags. (A flag with its initial direction in the subspace would remain so under any transformation in $\mathfrak{L}_0$.)

We select once and for all the form $f = \sum a_{\rho\sigma} u^\rho u^\sigma$ at O and shall transfer this form to each point of $\mathscr{S}$ in the following manner: Let P be any point of $\mathscr{S}$ and T any element of $\mathfrak{G}$ such that $T(O) = P$. The form f is carried by T to an induced form in the tangent space at P; this induced form is independent of T since the most general transformation in $\mathfrak{G}$ which takes O to P is obtained by composing a transformation in $\mathfrak{G}_0$ with T. But the transformation in $\mathfrak{G}_0$ induces an element of $\mathfrak{L}_0$ and this in turn leaves f unchanged.

Thus we obtain a Riemannian metric $ds^2 = \sum g_{\rho\sigma}(x)\, dx^\rho\, dx^\sigma$ on the whole space $\mathscr{S}$, and this metric is invariant under $\mathfrak{G}$. For if $T \in \mathfrak{G}$ and $T(P) = Q$, then we may obtain the passage $T: P \longrightarrow Q$ by a composition

$$P \longrightarrow O \longrightarrow Q$$

of transformations in $\mathfrak{G}$; under each of these we leave ds^2 unchanged.

Having this, let us consider the case $n = 2$. We have an invariant

$$ds^2 = g_{11}\, dx^1\, dx^1 + 2g_{12}\, dx^1\, dx^2 + g_{22}\, dx^2\, dx^2$$

on $\mathscr{S}$, and we consider the Gaussian curvature $K(x^1, x^2)$. Since $\mathfrak{G}$ is transitive on $\mathscr{S}$ and ds^2 is invariant under $\mathfrak{G}$, this curvature must be constant. Since we may multiply ds^2 by a positive constant, we may normalize to the following three cases

$$K = +1, \quad K = 0, \quad K = -1.$$

But this means that a neighborhood of a point on $\mathscr{S}$ is isometric to a neighborhood of spherical space, Euclidean space, or Bolyai-Lobachevski space, respectively; hence, in the small, the group $\mathfrak{G}$ is what we asserted it to be.

Now we return to the n-dimensional case and shall prove that locally the space $\mathscr{S}$ has the structure of the sphere $\mathscr{S}^n$, Euclidean $\mathscr{E}^n$, or the Bolyai-Lobachevski $\mathfrak{L}^n$. We take a plane element through a point O of $\mathscr{S}$ and form the surface spanned by all geodesics through O tangent to the plane element. The Gaussian curvature of this surface at O is the **Riemannian sectional curvature** K. This however is constant for $\mathscr{S}$, independent of O and the plane element, because any two plane elements at points of $\mathscr{S}$ are conjugate under $\mathscr{S}$. As before, we may make this constant $K = +1$, $K = 0$, or $K = -1$. One proves in differential geometry that locally the space $\mathscr{S}$ must be isometric to one of the three standard types.

We shall proceed by induction on the dimension n. We take a small number a and form the geodesic sphere of radius a about O by marking off the distance a on each geodesic from O. This is an $(n-1)$-dimensional space we shall denote by $\mathscr{S}^{n-1}(a)$. Now we consider the action of $\mathfrak{G}_0$ on this space. Since $\mathfrak{G}_0$ leaves O invariant and since $\mathfrak{G}$ leaves geodesic lines and distances invariant, it follows that $\mathfrak{G}_0$ sends $\mathscr{S}^{n-1}(a)$ onto itself. Thus $\mathfrak{G}_0$ induces a group of transformations in $\mathscr{S}^{n-1}(a)$. We verify without difficulty that this group satisfies the postulates originally given for $\mathscr{S}$, but in one lower dimension. We note that $\mathscr{S}^{n-1}(a)$ is compact and topologically equivalent to the $(n-1)$-sphere $\mathscr{S}^{n-1}$.

Suppose we have already proved in dimension $(n-1)$ that if a manifold $\mathscr{M}$ satisfies our axioms and in addition is homeomorphic to $\mathscr{S}^{n-1}$, then the geometry must be of the Riemannian spherical type. Then it would follow that this is the case for $\mathscr{S}^{n-1}(a)$. Since each $\mathscr{S}^{n-1}(a)$ is orthogonal to each geodesic through O, it follows that in a neighborhood of O we have

$$ds^2 = da^2 + \phi(a) \, (ds^*)^2,$$

where $\phi(a)$ is a positive factor, constant on each sphere, and ds^* is the line element of the sphere $\mathscr{S}^{n-1}$ induced by $\mathscr{E}^n$. It remains to determine ϕ. This is done by using the constancy of the Riemann curvature, a condition which leads to a second order differential equation for ϕ. By solving this equation we find only the three standard cases.

We have shown that our space $\mathscr{S}$ has the desired structure locally. It remains to pass to the corresponding result in the large.

Example. The torus obtained by identifying opposite sides of a period parallelogram in the Euclidean plane has local Euclidean structure, but is not similar to $\mathscr{E}^2$. We shall see that such spaces cannot satisfy our hypotheses.

For simplicity we shall assume in the next few paragraphs that the curvature $K = 0$, although the proofs for the cases $K = +1$, $K = -1$ are practically identical.

Each point of $\mathscr{S}$ has a neighborhood which is isometric to the interior of the Euclidean sphere of radius ρ. Since $\mathfrak{G}$ acts transitively and isometrically on $\mathscr{S}$, we may select the same constant ρ for *all* points of $\mathscr{S}$. We begin with a definite such isometry in a neighborhood of a point O of $\mathscr{S}$ and shall prove that it can be extended (by a process like analytic continuation) to all of $\mathscr{S}$.

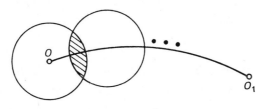

FIGURE 8

Recalling that $\mathscr{S}$ is assumed connected, we may connect O to any other point O_1 by an arc. We subdivide the arc by a finite number of points so that the corresponding neighborhoods overlap. The mappings in the overlap of the first two regions (shaded in Figure 8) differ only by a constant Euclidean isometry. This means we may change the second mapping so that the two mappings in the first overlap coincide. We proceed in this way until we arrive at O_1. The resulting mapping at O_1 is independent of the subdivision of the arc (as in analytic continuation); also two sufficiently close arcs joining O to O_1 yield the same result. Hence two curves joining O to O_1 which may be deformed into one another yield the same extension. We consequently have the following result.

LEMMA 38.1. *If $\mathscr{S}$ is simply connected, then there exists an isometric mapping on $\mathscr{S}$ into $\mathscr{E}^n$. It follows that the metric on $\mathscr{S}$ is given by*

$$ds^2 = \sum_{\rho,\sigma} g_{\rho\sigma}\, dx^\rho\, dx^\sigma = \sum_\tau (ds^\tau)^2.$$

We next prove that this mapping is one-one and fills $\mathscr{E}^n$. To do this we go through the same steps in the reverse direction. We begin with an isometric mapping on the interior Ω of a sphere of radius ρ in $\mathscr{E}^n$ onto a neighborhood Ω' of $\mathscr{S}$. This mapping may be continued to each point of $\mathscr{E}^n$ since the radius ρ is uniform for all points of $\mathscr{S}$. Since $\mathscr{E}^n$ is simply connected, the mapping is uniquely defined on all of $\mathscr{E}^n$ into $\mathscr{S}$. But it is clear that this mapping and the previously constructed one on $\mathscr{S}$ into $\mathscr{E}^n$ are inverses of each other, hence the latter is one-one on $\mathscr{S}$ onto $\mathscr{E}^n$. We obtain the result:

THEOREM 38.2. *If $K = 0$ and $\mathscr{S}$ is simply connected, then $\mathscr{S}$ is topologically and metrically equivalent to $\mathscr{E}^n$, and $\mathfrak{G}$ is similar to the group of proper Euclidean motions.*

The cases in which $\mathscr{S}$ is simply connected and $K = +1$ or $K = -1$ are disposed of in exactly the same way. It now remains to remove the hypothesis of simple connectedness.

Thus let $\mathscr{S}$ be a space which satisfies our basic hypotheses, but is possibly multiply connected. We introduce the universal covering space $\overline{\mathscr{S}}$ of $\mathscr{S}$ together with the natural projection

$$p \colon \overline{\mathscr{S}} \longrightarrow \mathscr{S}.$$

This mapping p is one-one in the small. Also any curve on $\mathscr{S}$ may be lifted by p to $\overline{\mathscr{S}}$. This means that if γ is a curve on $\mathscr{S}$ with initial point P and if $p(\bar{P}) = P$ for some point $\bar{P}$ on $\overline{\mathscr{S}}$, then there is a unique curve $\bar{\gamma}$ on $\overline{\mathscr{S}}$ with initial point $\bar{P}$ such that $p(\bar{\gamma}) = \gamma$.

We shall now construct a group $\overline{\mathfrak{G}}$ of transformations of $\overline{\mathscr{S}}$ which will satisfy the same hypotheses as $\mathfrak{G}$ does on $\mathscr{S}$. We begin with the following lemma.

LEMMA 38.3. *Let $T \in \mathfrak{G}$, and let $\bar{P}, \bar{Q} \in \overline{\mathscr{S}}$. Assume that $T[p(\bar{P})] = p(\bar{Q})$. Then there exists a unique transformation $\bar{T}$ of $\overline{\mathscr{S}}$ such that*

1. $\bar{T}(\bar{P}) = \bar{Q}$,
2. *For each $\bar{A} \in \overline{\mathscr{S}}$, we have $p[\bar{T}(\bar{A})] = T[p(\bar{A})]$.*

Proof. The transformation $\bar{T}$ is constructed as follows. Let $\bar{A}$ be an arbitrary point in $\overline{\mathscr{S}}$ and let $\bar{\gamma}$ be any curve in $\overline{\mathscr{S}}$ joining $\bar{P}$ to $\bar{A}$. (See Figure 9.) Then $\gamma = p(\bar{\gamma})$ is a curve in $\mathscr{S}$ joining $P = p(\bar{P})$ to $A = p(\bar{A})$. We apply the given T to γ to obtain a curve $\gamma' = T(\gamma)$ joining $Q = p(\bar{Q}) = T(P)$ to $T(A)$. By the lifting property, there exists a unique curve $\bar{\gamma}'$ on $\overline{\mathscr{S}}$ with initial point at $\bar{Q}$ such that $p(\bar{\gamma}') = \gamma'$. We call the endpoint of the curve $\bar{T}(\bar{A})$. That $\bar{T}(\bar{A})$ is independent of $\bar{\gamma}$ is immediate from the simple connectedness of $\overline{\mathscr{S}}$. Properties 1 and 2 are clear from the construction. Also $\bar{T}$ is a transformation since the same construction may be used to construct its inverse.

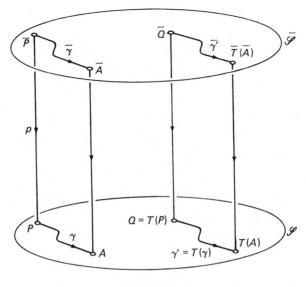

FIGURE 9

The totality of transformations $\bar{T}$ of $\overline{\mathscr{S}}$ constructed in this manner evidently forms a transformation group $\overline{\mathfrak{G}}$ of $\overline{\mathscr{S}}$. This group satisfies the basic axioms. For locally a neighborhood of $\bar{Q}$ is the same as a neighborhood of Q, hence the space of flags is the same. Also the transformation T may be replaced by the product of T and any transformation of $\mathfrak{G}$ which leaves Q fixed; thus we obtain the transitivity of $\overline{\mathfrak{G}}$ on the flags.

It follows by our previous results that $\overline{\mathfrak{G}}$ is similar to one of the three basic groups. To proceed further we must note the following fact about $\overline{\mathfrak{G}}$. If $\bar{P}_1$ and $\bar{P}_2$ are points of the universal covering space $\overline{\mathscr{S}}$, if $\bar{P}_1$ and $\bar{P}_2$ both cover the same point P of $\mathscr{S}$, and if $\bar{T} \in \overline{\mathfrak{G}}$, then $\bar{T}(\bar{P}_1)$ and $\bar{T}(\bar{P}_2)$ both cover the same point of $\mathscr{S}$. For

$$ p[\bar{T}(\bar{P}_1)] = T[p(\bar{P}_1)] = T(P) = T[p(\bar{P}_2)] = p[\bar{T}(\bar{P}_2)]. $$

We next introduce the discrete group of transformations on $\overline{\mathscr{S}}$ which consists of the sheet-interchanging mappings of $\overline{\mathscr{S}}$. This group leaves the ds^2 invariant. Suppose that $\bar{P}_1$ and $\bar{P}_2$ are points of $\overline{\mathscr{S}}$ conjugate under this discrete group. Then these points have the same projection onto $\mathscr{S}$, and consequently so do $\bar{T}(\bar{P}_1)$ and $\bar{T}(\bar{P}_2)$ whenever $\bar{T} \in \overline{\mathfrak{G}}$. This means that $\bar{T}(\bar{P}_1)$ and $\bar{T}(\bar{P}_2)$ are also conjugate under the discrete group. We consider three cases:

Case 1. $K = 0$. Then we have a discrete group $\mathfrak{H}$ of distance preserving transformations in $\mathscr{E}^n = \overline{\mathscr{S}}$ with the following property: if Q_1 and Q_2 are conjugate under $\mathfrak{H}$ and g is a proper Euclidean motion, then gQ_1 and gQ_2 are conjugate. It follows that if Q_1 is held fixed, then the orbit of Q_2 under the rotations about Q_1 must be discrete. This is clearly impossible unless Q_2 is forced to coincide with Q_1, or what is the same, $\mathfrak{H} = \{\epsilon\}$. In other words, $\overline{\mathscr{S}}$ is a one-sheeted covering of $\mathscr{S}$, consequently $\overline{\mathscr{S}} = \mathscr{S}$ is simply connected.

Case 2. $K = -1$. The same argument applies so that $\mathscr{S}$ must be simply connected, and hence is the geometry of Bolyai-Lobachevski.

Case 3. $K = 1$. In this case $\overline{\mathscr{S}}$ is the space $\mathscr{S}^n \subset \mathscr{E}^{n+1}$. If P_N is the north pole and P_S the south pole, then each motion on $\mathscr{S}^n$ leaving P_N fixed also leaves P_S fixed. Thus there is a two-element group, so either $\overline{\mathscr{S}} = \mathscr{S}$ or $\overline{\mathscr{S}}$ is a two-sheeted covering of $\mathscr{S}$. In this latter case, the motion group on $\overline{\mathscr{S}}$ is not necessarily transitive on the space of flags on $\mathscr{S}$. In fact, if n is even, there exist two transformations leaving a given flag fixed, but when n is odd everything is all right. Another way to see the difficulty in case n is even is this. The space $\mathscr{S}$ must be orientable, since fixing an orientation at a single point O of $\mathscr{S}$ determines the orientation everywhere by using the T in $\mathfrak{G}$. But

$$\mathscr{S}^n/\{\text{diametrically opposite}\} = \mathscr{P}^n = (\text{projective } n\text{-space})$$

is not orientable if n is even and is orientable if n is odd. This case of a two-sheeted covering we call **elliptic geometry**. It forms the single exceptional case to our results. It may be remarked that if we had gone up to n-dimensional flags and demanded simple transitivity in our hypotheses, we should have obtained all of the elliptic cases, not just those with n odd.

INDEX